Flavor Chemistry

ACS SYMPOSIUM SERIES **756**

Flavor Chemistry
Industrial and Academic Research

Sara J. Risch, EDITOR
Science By Design

Chi-Tang Ho, EDITOR
Rutgers State University of New Jersey

American Chemical Society, Washington DC

Library of Congress Cataloging-in-Publication Data

Flavor chemistry : industrial and academic research / Sara J. Risch, editor, Chi -Tang Ho, editor.

    p.   cm.—(ACS symposium series, ISSN 0097-6156 ; 756)

    Includes bibliographical references and index.

    ISBN 0–8412–3640–2

    1. Flavor—Congresses. 2. Flavoring essences—Congresses. 3. Food—Odor—Congresses.

I. Risch, Sara J., 1958– . II. Ho, Chi-Tang, 1944– . III. Series.

TP372.5 .F525   2000
664´.5—dc21   99-58097

The paper used in this publication meets the minimum requirements of American National Standard for Information Sciences—Permanence of Paper for Printed Library Materials, ANSI Z39.48–1984.

PRINTED IN THE UNITED STATES OF AMERICA

Foreword

The ACS Symposium Series was first published in 1974 to provide a mechanism for publishing symposia quickly in book form. The purpose of the series is to publish timely, comprehensive books developed from ACS sponsored symposia based on current scientific research. Occasionally, books are developed from symposia sponsored by other organizations when the topic is of keen interest to the chemistry audience.

Before agreeing to publish a book, the proposed table of contents is reviewed for appropriate and comprehensive coverage and for interest to the audience. Some papers may be excluded in order to better focus the book; others may be added to provide comprehensiveness. When appropriate, overview or introductory chapters are added. Drafts of chapters are peer-reviewed prior to final acceptance or rejection, and manuscripts are prepared in camera-ready format.

As a rule, only original research papers and original review papers are included in the volumes. Verbatim reproductions of previously published papers are not accepted.

ACS Books Department

Contents

INDEXES

Preface

The flavor of foods plays a critical role in consumer acceptability. Although sight of the food is the first impression that a consumer has, the initial aroma and the flavor once the product is eaten are important for a consumer to react positively to a product and want to eat that product again. Research is being conducted in a number of areas to understand the flavor of natural products and find ways to reproduce those flavors in processed food products. Consumers want not only convenience but also demand quality in products that are readily available.

The symposium on which this book is based was organized to present current research in the area of flavor chemistry. Two chapters present overviews of both academic and industrial research in the chemistry of flavors, flavor development, and flavor stability. It is difficult to completely cover the extensive research being conducted on the industrial side because much of this information is maintained as a trade secret. There are some areas that companies will apply for patents in, such as encapsulation; however, these are not revealed until after the patents issue. It can often be several years after the patents are applied for in the United States, which is well after commercial use has commenced. By the time the patents issue, the technology is no longer new to the marketplace, although the details are new to people not involved in the development of the technology. Only three chapters in the book are from the industrial side and the scope of these is somewhat limited by what the companies sponsoring them will allow to be printed.

On the academic side, researchers are looking into very specific areas, which will help us better understand what compounds are most important for the perception of a particular flavor, how they change, and the pathways to produce specific compounds that can contribute characteristic flavors to products. One of the interesting areas that is being researched is to understand which individual compounds in a particular flavor are most important to the consumer's perception of that flavor. This area of gas chromatography–olfactometry in which the compounds in a flavor are separated by gas chromatography and then detected by the human nose is providing insight into the compounds that seem to have the greatest impact on sensory perception. Out of several hundred compounds that might make up a natural flavor, not all are of equal importance to the human perception of the flavor. This methodology is being used to try to identify the most important compounds. This information can be used when trying to replicate the flavor of a product to be used in other applications.

Taking this idea one step further is to analyze the flavor compounds that are released when a person is chewing a product to compare to the aroma of the product that is sensed before consumption. One chapter addresses the research that is being conducted to try to understand flavor release while eating to fully recognize the compounds that the brain is sensing to give us the flavor that we identify with the product.

The research into flavors and flavor development continues to try to find new and better flavors for the consumer. This book addresses a variety of those areas to give an idea of the state of the art in flavor chemistry.

We thank all of the speakers who took the time to commit their presentations from a symposium into chapters for this book.

SARA J. RISCH
Science By Design
505 North Lake Shore Drive
Suite 3209
Chicago, IL 60611–3427

CHI-TANG HO
Department of Food Science
Rutgers State University of New Jersey
Cook Campus, 65 Dudley Road
New Brunswick, NJ 08901

OVERVIEW OF FLAVOR CHEMISTRY

Chapter 1

Trends in Industrial Flavor Research

Charles H. Manley

Takasago International Corporation, 4 Volvo Drive, Rockeligh, NJ 07647

The Flavor Industry has been going through a period of significant consolidation driven by the market forces to focus on major customers' research needs and economic pressures. Strategic partnerships have become the normal business practice of the industry. Such partnerships are based on the sharing of research efforts with the objective of making significant technical breakthroughs in new product development or solving product or process problems. The research focus for large flavor companies is to carry out their creative efforts as efficiently as possible and to tie their basic research efforts to market goals. The industry's efforts have been in the areas of analytical and synthetic chemistry, biotechnology, aroma component measurement, encapsulation, and addressing flavor problems of functional foods.

A major problem in dealing with flavors on international bases has been the patchwork of national regulations that control ingredients use. The basic concern is the safety of the ingredient at its intended use level. In the USA, we have in place the Generally Recognized As Safe (GRAS) concept. This concept has allowed the Industry to develop a recognized list of ingredients for use in food products. The Industry is now working to establish further scientific principles for evaluation of flavor ingredient safety so that an internationally acceptable list of ingredients may be established.

During the last few decades the cost of food to the American consumer has dropped significantly to a worldwide record low of about 10% of our income. Large grocery chains, discount chains and the desire to eat away from home have placed great pressures on the management of the Food Industry to produce high quality, safe food at the lowest possible cost. As suppliers to the Food and Beverage Industry the Flavor Industry faces similar challenges and pressures. This is the current business paradigm that has a significant effect on the direction of research in the industry.

The Partnership

There have been major changes in the structure of the Flavor Industry during the last 20 years. The industry has come a long way from an industry of many small custom fabricators of flavor, many privately owned by talented entrepreneurs, to the handful of consolidated multi-million dollar companies of today. Yet as these top companies were growing by consolidation of mid-sized companies, many small companies remained or were formed by the technical and sales entrepreneurs that made possible the industry of "old". So as the food industry has been changing, the restructuring of the Flavor Industry followed.

Part of the new paradigm is cost cutting and making the company more efficient. As a result the large international food and beverage companies want to deal with fewer suppliers for the obvious of economics. The scenario is to focus on a chosen few – "The Supplier List" - and for these few to form a partnership or "Strategic Alliance". To be accepted into the coveted group, a flavor company must have a history of business with the company and/or to be evaluated for their technology and business fits. In some cases, the uniqueness of a smaller flavor company will place it on the supplier list or allow it to be offered the prestigious "Strategic Partnership".

This new way of doing business has a marked effect on the research investments that the food, beverage or Flavor Company makes. This is the driving force for the trends in industry research.

The Business Arrangement

To make sure that a benefit is derived from the "strategic partnership" both parties agree on the purpose of the relationship. Basically the customer (Food and Beverage Industry) wants the supplier (Flavor Industry) to create new flavors, solve process and product problems, and deliver the flavors in timely fashion at the best (lowest) cost. From this partnership the supplier wants nearly exclusive right to participate in new product development, and a guarantee that it will be one of few companies or perhaps the only one to work on a project with a high sales potential. To work properly – the partners must be faithful to the needs of the other. Many of these partnerships are fairly new, so we must wait to see if, indeed, this new paradigm will succeed!

There is still a very healthy group of smaller flavor companies. However, the mid-sized companies have been nearly consumed in the consolidation phase of the industry that took place in the last 15 years. These smaller companies rely nearly completely on the flavor artistry of the "old" industry. Their business strategies are to create a flavor for a company where a good relationship exists – strong personal or professional contacts. They usually will deliver a small volume of product at a reasonable price. No high-tech investments either in people or equipment are needed, just the creativity of the flavorist in developing a flavor that works in the product. Perhaps you have noticed that some of the smaller food companies have become some of the most creative product innovators. They are the risk takers with new ideas.

The large companies are focused on finding ways to make their franchised products deliver more profit whereas the smaller companies are taking risks to launch new concepts in niche markets. You will find these comments as controversial as they are generalizations, but they are observations anyone in the industry can easily make.

Impact on Research

Now that there is some understanding of our business world, let us see what impact this has on research trends in the flavor industry. The first observation one can make is that the large food companies, which historically had flavor groups, have or are dissolving them. The long-term basic research in flavors done by these companies is being discontinued. We also find that less and less basic research is being done on flavor by major universities. This trend started many years ago, although one could believe that it should increase because of the needs for basic information by the business community. You will hear more about the academic side in the chapters to follow. Some thirty years ago there was major activity and publishing of research by many academic groups. That effort is now left to a few excellent groups, most of which are represented in this text. As noted, because of the direction of the business world, there are many flavor companies that are "compounders" only and do not have the resources to do "basic research." Those companies that are doing basic research are usually dedicated to the needs of their major "partners." Market trends also direct their research efforts. If you combine the needs of the big company partnership and the market trends you will find that the major areas of research on flavor are:

- Biotechnology
- Synthetic Organic Chemistry
- Encapsulation
- Aroma measurement techniques
- Process Flavors
- Flavor Ingredient Safety Evaluation
- Unique Food and Beverage Products

Some of these topics will be extensively reported on in this text, but a quick overview will be offered here.

Biotechnology

In the United States, our food laws require that any food that contains artificial components used as flavors be labeled as artificially flavored. This may appear on the principal display panel or in the ingredient declaration depending on the product's claims and the flavor added to the product. The FDA in the Code of Federal Regulation has defined "natural flavor" as "the essential oil, oleoresin, essence or extractive, protein hydrolysate, distillate or any product of roasting, heating or enzymolysis, which contains the flavoring constitutes derived from a spice, fruit juice, vegetable or vegetable juice, edible yeast, herb, bark, bud, root, leaf or similar plant material, meat, seafood, poultry, eggs, dairy products, or fermentation products thereof, whose significant function in food is flavor rather than nutrition"(1). Artificial ingredients are then defined as all other flavor ingredients not fitting this definition. The default position of no label claim is to consider a food product containing a flavor to be "naturally flavored." Therefore, if there are only two potentials for flavor labeling claims, the preferred labeling on most products is, of course, "natural." Most company's marketing groups see the term "artificial" as a negative when used on their products and request the product developer to use natural flavors.

The natural flavor definition has, therefore, initiated ways to produce flavor that fit the natural flavor definition. The use of enzymes (enzymolysis) and microorganisms (fermentation) has led the industry to commit to the use of "biotechnology" for the development of natural components. Both the use of enzymes and microorganisms have been employed for their use in converting natural food components (such as carbohydrates, protein, fats and vitamins) to flavor ingredients which have value as natural substances. You will read a great deal more about this area in the chapters to follow because it is a major research effort for both large and some small flavor companies. The natural flavor claim has prompted companies to have staffs containing microbiologists, biochemists and bioengineers to develop, scale up and commercialize flavor ingredients for use in foods and beverages. Table I indicates some of the microorganisms and their uses in creating flavor ingredients.

The major focus of research is in the area of biotechnology for the development of natural flavors has been –

- Isolation of interesting microorganism and enzymes for flavors use
- Safety evaluations and approval for use for the useful one
- Development of methods of manufacture
- Optimization of processes
- Genetic manipulation to express higher yield or purer products
- Development of methods of isolation and concentration of valuable components

Table I. Microorganisms used in flavor substance generation

Compound	Microorganism	Type
Methyl ketones	*Penicillum roqueforti*	Mold
6-Pentyl-α-pyrone	*Trichoderma viride*	Mold
1-Octene-3-ol	*Apergillus oryzae*	Mold
γ-Decalactone	*Candida lipolytica*	Yeast
Ethyl esters	*Geotrichum sp.*	Yeast
Propionic acid	*Propionibacteria*	Bacteria
Diacetyl	Lactic acid bacteria	Bacteria
Alkyl methoxypyrazine	*Pseudomonas perolens*	Bacteria

Table II gives examples of the cost of some of these naturally derived substances related to their synthetic analogs.

Table II. Comparative cost of naturally derived and synthetic flavor ingredient prices

Compounds	"Natural" price $/Kg	"Synthetic" price $/Kg
Acetaldehyde	135	47
Ethyl Acetate	7	1.60
δ- Decalactone	500	22
Cis-3-Hexenol	1325	67
Methyl nonyl ketone	385	53
1-Octene-3-ol	8250	102

Average prices as of 2/99

Biotechnology offers some of the most exciting elements of flavor research at this time. There has been a major shift from the organic synthetic chemist to develop these ingredients to the biochemist and microbiologist. However, the synthetic organic chemist still has a role to play in the Flavor Industry.

Chiral Chemistry

For a great many years flavorists have understood the flavor value and quality of certain materials due to their chirality or because of the specific structure (optical activity) of the chemicals involved. Certainly the use of natural Menthol is preferred to synthetic racemic Menthol and, so too, there is a significant aroma difference between the two isomers of Carvone (the l isomer gives a caraway note whereas the d isomer gives one a spearmint impression). Many flavor companies that are producers of synthetic chemicals have focused their synthetic research capabilities of this area. One of the first useful chiral sythetic flavor material, l-Menthol, was produced by Takasago International Corporation International using an allosteric catalysis method (2).

In the last 10 years or so, analytical chemists have developed excellent methods separating the chiral species of various natural chemicals. Various chiral separating phases have been developed for gas chromatography. Hydrogen bonding, coordination complexation and inclusion have been used. And advanced techniques of LC-GC and multidimensional gas chromatography (MDGC) have also been used. Figure 1 shows the enantiomer ratio for Citronella isolated from various natural sources (3). Although there are significant differences in the chiral properties of this component there is only a small aromatic difference. Linalool occurs in both enantiomeric forms in many products.

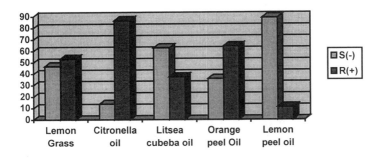

Figure 1. Distribution of enantiomers of Citronella from various natural sources (3).

The R(-) is present in sage whereas the S(-) Linalool is found in basil oil (4). You can see the significant differences in these materials, which relates to their usefulness as flavor ingredients. Flavorists know the value of each material for their creative efforts.

Today, many companies are synthesizing a variety of chiral compounds. Most of these substances are used as precursors for pharmaceuticals, but many are starting to show up in flavor and fragrances compounds. Examples of some of the substances available from the industry are shown in Table III.

As mentioned, the fuller understanding of the role of the specific structure of substances related to their occurrence in nature is due to advances in organic and analytical chemistry.

Table III. Chiral compounds used as flavor ingredients (Comparison of enatiomer odor thresholds)

Compound	Treshold
(S)-(-)-Carvone	2
(R)-(+)-Carvone	85-130
(R)-(+)-(E)-α- Damascone	100
(S)-(-)-(E)-α- Damascone	1.5
(+)-Nootkatone	800
(-)-Nootkatone	600,000

Innovations in Instrumental Analysis

Flavor companies are not the primary developers of instrumental methods for their research, but they have been significant contributors to unique applications and uses of the new technology. Chiral chemical separation is one example of the many areas of research. Other areas such as high performance liquid chromatography, micro-bore gas chromatography and electronic detection of odorants by the so-called "electronic nose" have also joined the flavorist "tool box".

The "electronic nose" represents an interesting approach to measuring aroma profiles. There are three types of "electronic noses" commercially available to the flavor industry today. The attributes of each are shown in Table IV. Although these devices are of interest to the industry, we have determined that they will not replace the human nose as an important detector in flavor research for some time to come (5).

Table IV. Comparison of "electronic nose" attributes

Company	Model	Type of Sensor	# of Sensors
Alpha M.O.S.	Fox 2000	Metal Oxides Conducting polymers	6, 12 or 18
AromaScan	Aromascanner	Conducting Polymers	32
Neotronics	The Nose	Conducting Polymers	12

A second type of aroma evaluation technique deals with the sniffing of the diluted effluent of a gas chromatograph. The two similar techniques "CharmAnalysis™" (6) and Aroma Extraction Dilution Analysis (AEDA) (7) are established methods that produce quantitative estimates of relative potency for the aroma compounds that elute from the gas chromatograph. The disadvantage of these methods is that several dilutions must be sniffed until no significant aroma can be detected. This is very time consuming. However, the reproducibility of this technique is very good (*8*) and its ability to provide an evaluator with a standard dose of a pure odorant is excellent.

Encapsulation

The major route for encapsulation of flavors has been by the use of botanical gums and/or carbohydrates (dextrin and/or modified starches) via a spraydrier process. Encapsulation by this method offers an economical way to render a convenient solid form of the liquid flavor. It also reduces the volatility of the product and offers some protection against oxidation (*9*)

A recently approved material, β cyclodextrin, is now being used as an entrapping (inclusion complexes) material prior to spray drying (*10*). The material is expensive and has only been approved for food use in 1998. Therefore, it has only found unique niche uses in the flavor industry at this time. The material is very effective in reducing the loss of flavor due to volatilization, oxidation or light reactivity (*11*). Flavors with high aroma content are also fixed so that very little or no aroma is perceived in the dry form.

Spray chilling or spray cooling is a method that uses cool air to set a fat coating around a flavor compound (*12*). This method is particularly good for encapsulation or water-soluble flavors.

Extrusion technology based on making a glass-like extrusion of sugar with a dextrin ingredient has found a place in commerce for encapsulation flavor oils (*13*). This method offers the best protect against oxidation and the extruded bits may be colored for visual effects. Major use of this encapsulated material has been in the confectionery and chewing gum industry. A recent advance making a non-sugar extrusion has been based on the use of sugar alcohol (*14*).

Process Flavors

The art of cooking has been reflected in the science of flavor creation in the area of process flavors. These flavors are developed by reacting various food components under thermal conditions to produce a profile similar to many types of cooked or roasted foods. The major reaction flavors have been created by the knowledge of the Mallard Reactions (non-enzymatic browning) which are based on amino acids reacting with reducing sugars. Other reactions such as decomposition of fats and oils and sugar dehydration also play an important role in the development of Process Flavors. Process Flavors range from nuts and chocolate to chicken and beef. Once again, the driving force in their development has been the definition of "natural" (*15*).

Natural food components thermally reacted to produce flavor results in the formation of a "natural flavor" per the CFR title 21 definition given previously. Although the first patent in this area dates to the early 1960's there is still a very significant amount of research being done in this area to refine the amounts and type of precursor materials and their processing conditions (*16*).

Flavor Ingredient Safety Evaluation

The flavor industry has always regarded the review of flavor ingredient safety as a primary focus of research for the industry. Under the industry's trade association, the Flavor and Extract Manufacturers' Association (FEMA) and the world group, the International Organization of Flavor Industries (IOFI), the Flavor Industry has established an independent Expert Panel for the review of flavor ingredient safety. The 1958 Food Additives Amendment to the Pure Food Act of the USA allows for ingredients used at very low levels or used historically to be approved for food use as materials which are Generally Recognized As Safe (GRAS). The Act requires that professionals qualified to make such food safety judgements make the safety reviews and approvals. This review may be made without the government's (FDA) review or approval and the material may be used as long as it is not a poison at its intended use level. FEMA established its GRAS list of allowed substances in 1965 (*17*) and since then has added more than 1800 substances to that list (*18*). The scientific methods used and publications issued by the Expert Panel and industry scientists has established the FEMA safety review process as a standard respected by the USA regulatory groups and by some 46 nations worldwide.

With the trend in globalization of the food and beverage industry, there is an urgent need to develop a harmonized list of ingredients that would be permitted in all countries. Currently FEMA is working with IOFI, the worldwide trade organization, to bring about that harmonization. IOFI has been working with member organizations to establish such a list of flavor ingredients it considers safe for use in food. An international scientific board has been established to provide safety data for use in the multi-national approval of substances. The European Union will establish their list of allowed substances by 2003 (*19*). It is expected that with harmonization between the US and Europe the rest of the world will accept both the ingredients and the scientific review methods used for safety evaluation as appropriate for use in their countries.

These break-throughs in harmonization of flavor ingredients will open up the potential for new ingredients to be developed. Currently a company that achieves GRAS in the USA for a substance still needs to move ahead to qualify the flavor ingredient in many different countries, some with very small market potentials. However, with an established way to get general safety approval in many countries, because of their acceptance of the scientific principles of the safety review, the market opportunity will drive companies to invest in research to develop more interesting and novel flavor substances.

The FEMA Expert panel will be involved in the review of approximately 600 ingredients allowed in Europe and the European authorities will be reviewing FEMA GRAS substances for addition to their inventory. The results of these reviews will allow a global food or beverage company to use a flavor compound to be used across many countries. This is truly a major future trend and driving force in the flavor industry.

Unique Food and Beverage Products

The food and beverage marketing trends have a great influence on the direction of research in the Flavor Industry. Its influence may be characterized as having the strongest effect on shifting the creative direction. For example, a trend in ethnic foods will channel flavorists to use their creative skills in compounding flavors more appropriate for those particular markets, i.e. hot and pungent flavors for a Latin line of products or savory type for an oriental line of products.

One of the developing markets, which will and has required some basic research, has been the evolving nutraceutical market. Reliance on ingredients with known health value does not necessarily make the final product a palatable one. When the health market was mainly focused on people who were concerned for their health only, the flavor or organoleptic issue was not a major one. However, as the concept has gone mainstream the products must be acceptable organoleptically to a mass market in order to survive. They must have a pleasant flavor and taste.

Research into the compatibility of flavors and nutraceutical ingredients is one which is coming into its own. The industry is learning to cover, attenuate or flavor around the off-flavor and taste notes associated with so many of the nutraceutical ingredients. With the aging of the industrialized world there is a trend for people to want "functional" foods and beverages that can deliver a health claim or claims and taste good (20)!

A second area of flavor research is reacting to the needs of the Food Service Industry. This segment of the Food Industry is fast becoming "the" major supplier of food to the people of developed countries. The stability of quality flavor through major distribution channels and delivery methods has required flavor companies to think differently about their flavors and to make significant use of their applications groups (21).

Conclusions

The flavor industry is following the food and beverage industry through a period of consolidation. This consolidation has led to the development of "strategic partnerships" between the two industries followed by a dedication of research effort between the partners. This business trend plays a major force in the direction of industrial flavor research. Although there are smaller companies where product innovation is taking place, the major research direction is coming from the strategic partnerships.

This text will relate, in detail, some of the fascinating research being done by the Flavor Industry and Universities. The Flavor Industry remains a dynamic force in the field of flavors and aroma research, as many companies spend a significant part of their income developing new substances, processes and flavors for the Food and Beverage Industry of the world.

Literature Cited

1. Code of Federal Regulations, Title 21, Section 101.22. US Publication, Washington, D.C.
2. Kumobayashi, H. *Reci. Trav. Chim. Pays-Bas* **1996**, vol. 115; 201-210.
3. Yamamoto, T. *Koryo* **1994**, vol. 184; 57-72.
4. Boelens, M. H. 1993. Sensory properties of optical isomers. *Perfumer and Flavorist.* **1993**, vol 18, 2-14.
5. Freund, M. S. and Lewis, N. S. *Proc. Natl. Acad. Sci.* U. S. A. **1995**, vol. 92(7), 2652-2656.
6. Acree, T. A.; Barnard, J. and Cummingham, D. *Food Chem.*, **1984,** *vol. 14*: 273-286.
7. Ullrich, F. and Grosch, W. *Z. Lebensm. Unters. Forsch.* **1987**, vol. 184(4); 277-282.
8. Acree, T. A. In *Flavor Measurement;* Ho, C-T. and Manley, C. H.; Editors; Marcell Dekker Pub., New York, NY, 1993: pp77-94.
9. Reineccius, G. A. *Food Review International* **1987**, vol. 5; 147-176.
10. Jackson, L. S. and Lee, K. *Lebensm. Wissn. Technol.* **1991**, vol. 24; 289-297.
11. Pagington, J. S. *Food Flavors, Ingredients, Packaging and Processing* **1985**, vol. 7; 51-55.
12. Lamb, R. 1987. Spray Chilling. *Food Flavors, Ingredients, Packaging and Processing* **1987**, Vol. 9; 39,41,43.
13. Reineccius, G. A. *Food Review International* **1987**, vol. 5; 147-176.
14. Tanaka, S., Manley, C. H. and Nagano, K. U. S. Patent 5,709,895, 1997.
15. Izzo, H. V.; Yu, T. H. and Ho, C-T., In *Prog. Flavour Precursors Stud. Proc. Int. Conf.*, Scheier, P., and Winterhalter, P. Ediotrs; Allured Publishing. Carol Stream, IL., 1993, pp 315-328.
16. May, C. G. British Patent 858,333, 1961.
17. Hallagan, J. B. and Hall, R. L. *Toxic. And Pharma.* **1995**, vol. 211; 422-430.
18. Emerson, J. L. and Stone, C. T. In *Flavor Measurement*; Ho, C-T and Manley, C. H.; Editors; Marcel Dekker Pub. New York, NY, 1993; pp 359-372.
19. Regulation (EC) No. 2232/96 of the European Parliament of the Council 28 Oct. 1996. *Official Journal of the European Communities* 23.11.96 No L 299/1.
20. Flesch, R. and Rychlik, K. *Food Processing* **1997**, 51,53,54.
21. Scheiber, W. L., Scharpt, L. G. and Katz, I. *ChemTech.* **1997**, March; 58-62.

Chapter 2

Recent Developments in Academic Flavor Research

Gary A. Reineccius

Department of Food Science and Nutrition, University of Minnesota,
1334 Eckles Avenue, St. Paul, MN 55108

This manuscript provides a brief historical perspective of the
driving forces that have motivated flavor research and then goes on
to present an overview of current developments in this field. The
key topics discussed include determining key aroma constituents of
foods, factors influencing aroma release from foods, "electronic
noses", thermally generated flavor, biotechnology to produce
flavors and lastly, a look into the future of flavor research in
academia.

Introduction

Progress in flavor research has been an evolutionary process. From a historical view,
flavor research was significantly driven by advances in instrumentation. Great strides
were made when gas chromatography became generally available (very late 50s to
early 60s). Prior to gas chromatography, the isolation, separation and identification of
unknown volatile compounds was an extremely tedious task. Gas chromatography,
even in its most primitive state, represented a spectacular step forward in flavor
chemistry. As gas chromatography evolved in sophistication, so followed progress in
flavor chemistry. The advent of fused silica capillary gas chromatography columns
was particularly significant since fused silica column development did not limit high
resolution chromatography to a hand full of experts but made it possible for all.

The development of low cost quadrapole mass spectrometers also has resulted in
significant advances in flavor research. Low cost instruments with excellent GC
compatibility has also put this technique in the hands of many flavor researchers who
otherwise could not afford the technique. Unfortunately, this development has also
been a curse in some ways in that it is occasionally used by researchers who do not
adequately confirm compound identities and erroneous identifications enter the
literature.

Beyond instrumental developments, flavor chemistry has evolved in terms of understanding. Initially, researchers used GC/MS to identify long lists of aroma chemicals in foods. This has resulted in nearly 7,000 aroma compounds identified in foods today (1). Many of these aroma compounds are present naturally in foods while others are the result of fermentation, thermal processing or deteriorative reactions (e.g. lipid oxidation). It was noted relatively soon that food flavors could not be regenerated from these lists and some logical approach had to be formulated to determine which aroma compounds made a significant contribution to food aroma and which were insignificant. The earliest attempts in this area were to determine the sensory character of individual aroma compounds as they eluted from a GC (GC Olfactometry). Those aroma compounds that smelled like the food were considered most important. Unfortunately, many foods did not contain "character impact compounds" but the aroma was the result of a combination of numerous noncharacteristic odorants. This issue had to be addressed differently and has resulted in numerous related techniques for determining the key aroma constituents of foods (2, 3, 4). The earliest technique was that of simply determining if an odorant was present in a food was above its sensory threshold. Rothe and Thomas (5) added a quantitative aspect to this by calculating the odor values of aroma constituents in a food – essentially concentration of an odorant divided by its sensory threshold in a food. While this technique has developed by several researchers (6-9), conceptually it has come under considerable criticism and work remains to be done in this field as is discussed later in this paper.

Historically, considerable effort has been devoted to identifying mechanisms of flavor formation in plants (biosynthesis), during heating (Maillard reaction), and fermentation. Off flavors have been a topic of considerable study as well due to the economic significance of this area to the food industry (10). Studies on mechanisms of flavor formation are waning due to changes in the funding of flavor research throughout the world. The effect of these changes on academic research are discussed at the end of this paper.

Current Developments

I have to admit at the outset of this section that I will be presenting current efforts in the field from my vantage point and knowledge base. This will unfortunately leave out some very valuable research due to my interests or oversight. I apologize for these omissions in advance.

Determining Key Aroma Constituents of Foods

One of the objectives of flavor research, industrial or academic, is to identify key aroma components of a food. Researchers in the flavor industry may choose to do this in order to determine the aroma components required to formulate a natural or artificial flavor. The food industry may pursue similar objectives but for quite different purposes. A food company may want to determine the key aroma components of a product to understand how changes in processing, formulation or packaging will impact flavor. The value of this point can be made better by example. If we are interested in making a product with a longer shelf-life (assuming that flavor limits shelf-life), it is useful to know the mode of flavor failure. Are we losing desirable (key) aroma components allowing off notes to surface and be detected or are we retaining the desirable aroma compounds but off flavors are forming and masking the desirable notes? If we are losing the desirable flavor notes, how are we losing them – to oxidation, interactions with the food itself, or other mechanisms of flavor loss? It is difficult to imagine how one is to develop methods (e.g. processes or packaging) to protect the flavor of a food if one does not know what aroma components one is to protect and from what. Yet, very few food companies have invested the resources to analytically characterize the flavor of their products.

The process of characterizing the key aroma constituents of a food have evolved greatly since the early sniffing work (2, 11). However, there still are major problems associated with the methodologies (11-14). One major problem with all of the current approaches is that they attempt to evaluate the contribution of a given odorant to a complex flavor totally out of context i.e. typically separately from all other aroma components and out of the food matrix. These approaches involve isolating the aroma of from a food, separating the aroma into components on a GC and then using dilution, intensity or frequency of sensing to determine importance. With time, it has become recognized that none of these methods reliably determine key aroma components but are screening methods that suggest key aroma components. Inevitably, sensory work must follow to evaluate the qualitative and quantitative data obtained. Fortunately, more sensory work is being done to validate the method results although some of this work is being done in laboratories ill trained or equipped for sensory studies.

Additional research is needed to lend more strength to the methods being used to select key aroma components. For example, there are no guidelines for the number of aroma components to select or basis for selection of these aroma components. The observation that they are present at the highest dilution factor or frequency of sensing may not be the most rational criteria. An aroma component may make a significant sensory impression at very low dilution (or frequency) if it is very obnoxious. Also, some aroma components may never make a significant contribution due to their low sensory intensity even at high dilution factors (or frequency). Thus, we need to have more sensory work done relating sensory response to dilution factors (or other selection method) and mixture work to better understand the criteria for an odorant changing the character of a mixture. We might also look at new approaches. One approach might be to prepare an aroma isolate of a food by several techniques and then judge the isolates for authenticity. Choosing the most authentic isolate would mean that all odorants needed to reproduce an aroma are present in the isolate and in the proper proportions. These two points are significant since obtaining pure aroma components and then deciding on concentrations for sensory evaluation are problematic. Through fraction collection from capillary columns, one might then select odorants from a GC run on any number of criteria and using collection and recombination, ultimately match sensory profile. An advantage is that the contribution of an odorant would be judged in a mixture as opposed to individually. Only after determining what components are needed to reformulate an odor does one have to actually do identification work and source pure components for further work. The primary weakness of this approach is that capillary columns do not yield significant amounts of material for sensory work. Thus, multiple GC runs are needed to collect sufficient amounts of material for sensory evaluation. This can be very tedious.

Aroma Release

If one considers what is required to give a sensory response, first one must have the needed aroma compounds (key aroma compounds) and secondly, they must be released from the food. If either is missing or out of balance, the flavor of the food will be incorrect. The importance of flavor release is obvious when one considers a low calorie food produced either through the use of high intensity sweeteners or reduced fat content. One can use exactly the same flavor and find that it is very acceptable in a full calorie food but quite unacceptable when used in a low calorie version of the product. Since the same aroma compounds (and concentrations) are present in the two products, the difference in the sensory properties of the products is the result of different flavor release from the foods.

A detailed discussion of flavor release from foods during eating has been provided by Taylor (15) and Haring (16). Taylor (15) has summarized some of the factors influencing release as:

- textural properties of the food including gel strength and viscosity;
- binding to major food constituents including proteins and starch which result in vapor pressure lowering;
- solubilization by fat; and
- interactions with minor constituents such as aspartame (Shiff base formation   with aldehydes)
- rehydration of a dry food
- chewing
- enzymes in either the food or mouth

The role of flavor interactions with major food constituents, e.g. starch and protein, in influencing flavor release has been researched extensively in the US in the 70s and 80s (17) . There is some work continuing in the US today but it is very limited in scope (18-20). However, this area is being studied intensely in Europe as is evidenced by the number of papers presented at the last Weurman Symposium (21). A substantial effort has dealt with methods to measure aroma release from foods (22). It is evident that one can not do much to study flavor release from foods if one can not measure it. Early studies considered vapor pressure reductions due to flavor binding (most of early US work). However, this technique does not consider the dynamic effects of texture in limiting flavor release. Dynamic methods were developed employing purge and trap methodology and ultimately "artificial mouths" were developed to simulate chewing (23-25). These artificial mouths were often quite simple devices based on a blender to provide controlled shear, temperature control to hold at body temperature and then some air flow to sample what is released from the food. Most of these methods lacked sufficient sensitivity to collect data in real time and therefore, a concentration step was required. Collection times were often in minutes while in real life one seldom keeps food in the mouth more than a few seconds. Concern was expressed that in some cases critical factors may have been missed due to the long time frame required by the analytical procedure. For example, the temporal profile of flavor release may influence sensory perception.

Andrew Taylor (22) was the first to develop a real time method for measuring aroma release in the mouth. Over time he has refined this method and been publishing on its application (21). Professor Taylor has a chapter in this book on his methodology and its application so this topic will not be pursued here. In my opinion, this is one of the most significant developments in the flavor area in recent time. Prof. Taylor has given us the tools to accurately evaluate theories relating food flavor interactions and sensory response.

"Electronic Noses"

"Electronic noses" have been the subject of considerable research in the US and Europe (26-27). When I first heard of "electronic noses" I have to admit I was quite excited. I gave a paper at the ACS extolling their potential applications (28). In theory, they offer exactly what we need to address some of our quality control issues, geographical origin of products and perhaps even some predictive work (shelf-life). As time has progressed, I have become disenchanted with the tool and very wary. My primary concern relates to the fact that there is no known (or understood) basis for instrument response. When a detector array response pattern is generated, we do not have any basis for understanding what was detected. The detector array may have responded to what we wanted to measure or something unrelated and potentially erroneous. For example, a researcher presented a paper at the 1996 IFT on using the electronic nose to measure oxidation in meats during frozen storage. He subjected the samples to the electronic nose and to a sensory panel. The sensory panel was asked to determine the level of oxidized off flavor. As one would expect, the electronic nose software established a correlation between some detector array response and the sensory panel. The electronic nose found something changing during storage and the sensory panel found the samples to be increasing in oxidized off flavor. However, one has no assurance that the electronic nose was responding to oxidized flavor. If the researcher had asked the sensory panel to judge color, Maillard off flavor or even moisture content, the electronic nose would have developed a correlation to that parameter instead. If two things are changing, a correlation may be found. This correlation was preselected by the researcher. Thus, without an assurance that the electronic nose is responding to what we want to measure, it can not be relied upon. Some studies have been done to determine what sensors respond to what chemicals but this is generally done as individual aroma compounds or simple mixtures. We have no assurance of what will happen in complex mixtures. I do not question that very valid and useful applications will be found for the electronic noses. I just suggest caution in accepting the literature and results until well proven.

A recent development in this area is the Chemical Sensor offered by Hewlett Packard. This instrument is based on obtaining a complete mass spectrum (no separation) of the volatiles in the air and then using chemometrics to establish correlations between the total spectra and sensory panel judgements. While this instrument can also draw erroneous correlations, there is an understandable relationship between the data and response. For example, one can envision that a musty grain sample would give MS ions characteristic (unique?) of isoborneol or geosmin and thus, give a response of musty off flavor in the grain when these volatiles are present in the product. Or, based on understanding that the shelf-life of fluid milk depends upon previous microbial growth (before pasteurization), one could understand the Chemical Sensor giving a predicted shelf-life based on the amount of acetone or alcohols (microbial metabolites) in the milk at the time of pasteurization. This may permit the milk bottler to screen the milk for off-flavor and put a useful shelf-life dating on the carton.

Thermally Generated Flavor

There has been a long-term interest in studying the development of flavor via thermal processing (29-31). This can be in foods as a part of normal processing or through the use of reaction systems to produce flavorings. This research has lead to a limited understanding of the mechanisms and precursors of flavor formation through reactions such as the Maillard reaction and means to control these reactions. One of the frustrations in this area has been the complexity of the reactions and their acute sensitivity to minor changes in formulation or processing conditions. Thus, much of the practical work done in controlling flavor during such reactions is empirical in nature.

Science has made a contribution in this area as is evidenced by Schieberle's work on the formation of 2-acetyl-1-pyrroline (32, 33). If one wants to enhance the formation of this particular bread crust, cracker or popcorn note in a food, his work on precursors and conditions for formation has provided an invaluable knowledge base. One might also suggest as examples, much of the work of Ho (general reactions and flavor formation, 34-36), Farmer and Mottram (meat-like flavor, 37-39), and Rizzi (pyrazine work, 40, 41) as being in this same category (this list is not all inclusive but an example of the literature).

Biotechnology to produce flavors

A limited amount of research in this area continues to be done in academic institutions (42-44). The vast majority of work is done in industrial settings with the goal of producing natural aroma compounds for flavor formulation (45,46). While academic institutions had substantial research programs in this area in the 80's and early 90's, only a few research institutions maintain strong programs in this area today.

Future Flavor Research in Academia

I would like to address two issues here. The first is that there is little question that flavor research in the US and Europe is becoming much more applied in nature. This is the result of cutbacks in federally funded research programs in the flavor area. When funding is limited, it is difficult for (even) me to argue that flavor is more important to society than food safety or nutritional well being. The outcome is that more universities are funded by the food or flavor industries. We will continue to see problem solving being done by universities with limited basic research to build upon in the future. A very disconcerting aspect of this shift in funding is the affect it is having on the free presentation and discussion of results within academic settings. It is a sad state to find that university professors can not present or discuss their work with each other due to confidentiality or patent constraints.

A second is that there is little or no coordination of flavor research efforts in the US. The European community has provided the forum and financial means to gather researchers in important topical areas (e.g. flavor release). In some cases, there are no actual funds given to support research (COST program) but funds are provided to facilitate yearly (or more frequent) meetings between all researchers who have active research programs in a given area. When funding is so limited, it is beneficial that there be effective coordination of efforts to best use resources. The European program should serve as a model for us in the US. Minimally, it would serve us well to meet formally at a national meeting to discuss and coordinate efforts to broadly enhance funding of flavor research and coordinate our research programs.

Bibliography

1. TNO-CIVO Food Analysis Institute. *Volatile Compounds in Food.* Utrechtseweg, The Netherlands, 1995.
2. Blank, I. In *Techniques for Analyzing Food Aroma.* R. Marsili, Ed.; Marcel Dekker Inc.: New York, 1997; pp. 293-330.
3. Mistry, B. S.; Reineccius, T.; Olson, L. In *Techniques for Analyzing Food Aroma.* R. Marsili, Ed.; Marcel Dekker Inc.: New York, 1997; pp. 265-292.
4. Grosch, W. *Trends in Food Science and Technol.* 1993, 4, 68.
5. Rothe, M.; Thomas, B. *Z. Lebensm. Unters. Forsch.* **1962**, 119, 302.
6. Grosch, W. *Trends in Food Sci. Technol.* **1993**, 41, 68.
7. Acree, T.E.; Barnard, J.; Cunningham, D.G. *Food Chem.* **1984**, 14, 273.
8. McDaniel, M.R.; Miranda-Lopez, R.; Watson, B.T.; Micheals, N.J. Libbey, L.M. In *Flavors and Off-Flavors.* G. Charlambous, Ed., Elsevier Publ.: Amsterdam, The Netherlands, 1990; p. 23
9. Ott, A., Montigon, F., Baumgartner, M., Murioz, R., Chaintreau, A. *J. Agric. Food Chem.* **1997,** 45, 2830.
10. Saxby, M.J. *Food Taints and Off-Flavors.* Blackie Academic & Professional: London, 1996 (2nd edition); 326.
11. Mistry, B.S.; Reineccius, T.; Olson, L.K. In *Techniques for Analyzing Food Aroma.* R. Marsili, ed.; Marcel Dekker Inc.: New York, 1997; pp. 265-292.

12. Frijters, J.E.R. *Chem. Senses Flav.* **1978**, 3, 227.
13. Etievant, P.; Moio, L.; Guichard, E.; Langois, I.; Schlich, P.; Chambellant, E. In *Trends in Flavor Research.* H. Maarse; G. van der Heij, Eds. Elsevier:Amsterdam, 1994; p. 179.
14. Abbot, N,; Etievant, P.X.; Issanchou, S.; Langois, D. *J. Agric. Food Chem.* **1993**, 41, 1698.
15. Taylor, A.J. In: *Critical Reviews in Food Science and Nutrition.* CRC Press: Cleveland,OH, 1996; p. 765.
16. Haring, P.G.M. In *Flavor Science and Technology.* Y. Bessiere; A.F. Thomas, Eds. John Wiley & Sons: Chichester, 1990; p. 351.
17. Schirle-Keller, J.P. *Flavor Interaction with Fat Replacers and Aspartame.* Ph.D. Thesis. 1995; University of Minnesota, St. Paul, MN
18. Schirle-Keller, J. P.; Reineccius, G. A.; Chang, H. H. *J. Food Science.* **1992**, 57(6), 1448.
19. Schirle-Keller, J.P.; Reineccius, G.A.; Hatchwell, L.C. In *Food Flavor Interactions.* R.C. McGorrin and J. Leland, Eds. *American Chemical Society Symposium Series.* American Chemical Society:Washington D.C. 1996; p. 143.
20. McGorrin, R.C.; Leland, J. (Eds). *Food Flavor Interactions.* American Chemical Society Symposium Series #633, American Chemical Society:Washington D.C. 1996.
21. Taylor, A.J.; Mottram, D.S. *Flavour Science: Recent Developments.* Royal Society of Chemistry:London, 1996; p. 476.
22. Taylor, A.J.; Lindforth, R.S.T. In *Trends in Flavor Science.* H. Maarse; G. van der Heij, Eds. Elsevier:Amsterdam. 1994; p. 3.
23. Lee, W.E. *J. Food Science* **1986**, 51, 249.
24. van Ruth, S.M.; Roozen, J.P.; Cozijsen, J.L. In *Trends in Flavor Science.* H. Maarse; G.; van der Heij, Eds. Elsevier:Amsterdam. 1994; p. 59.
25. Roberts, D.D. *Flavor Release Analysis using a retronasal aroma simulator.* Ph.D. Dissertation, Cornell University, New York, 1996.
26. Talou, T.; Sanchez, J.M.; Bourrounet, B. In *Flavour Science: Recent Developments.* Royal Society of Chemistry:London, 1996; p. 277.
27. Talou, T.; Maurel, S.; Gaset, A. In *Food Flavors: Formation, Analysis and Packaging Interactions.* E.T. Contis, C.T. Ho, C.J. Mussinan, T.H. Parliament, F. Shahidi, A.M. Spanier, Eds. Elsevier Publ:Amsterdam, 1998; p. 79.
28. Reineccius, G.A. In *Chemical Markers for the Quality of Processed and Stored Foods,* T.C. Lee and H.J. Kim, Eds. ACS Symposium Series #631, American Chemical Society:Washington D.C. 1996; p. 241.
29. Parliament, T.H.; Morello, M.J.; McGorrin, R.J. *Thermally Generated Flavors.* ACS Symposium Series #543, American Chemical Society:Washington D.C. 1994.
30. Parliament, T.H. ; McGorrin, R.J. *Thermal Generation of Flavors.* ACS Symposium Series #409, American Chemical Society:Washington D.C. 1989.
31. Labuza, T.P.; Reineccius, G.A.; Monnier, V.; O'Brien, J.O.; Baines, J.W. *Maillard Reactions in Chemistry, Food and Health.* Royal Society of Chemistry: London. 1994; p. 440.
32. Schieberle, P. *Z. Lebensm. Unters. Forsch.* **1990**, 191, 206.
33. Hoffman, T.; Schieberle, P. *J. Agric. Food Chem.* **1998, 46,** 2270.

34. Yu, T.H.; Chen, B.R.; Lin, L.Y. Ho,; C.-T. In *Food Flavors: Formation, Analysis and Packaging Interactions.* E.T. Contis, C.T. Ho, C.J. Mussinan, T.H. Parliament, F. Shahidi, A.M. Spanier, Eds. Elsevier Publ: Amsterdam, 1998; p. 493.

35. Tai, C.-Y.; Ho, C.-T. *J. Agric. Food Chem.* **1997**, 45, 3586.

36. Zheng, Y.; Brown, S.; Walter, L.O.; Mussinan, C.J.; Ho, C.-T. *J. Agric. Food Chem.* **1997**, 45, 894.

37. Farmer, L.J.; Hagan, T.D.J.; Paraskevas, O. In *Flavour Science: Recent Developments.* A.J. Taylor; D.S. Mottram, Eds. Royal Society of Chemistry:London, 1996; p. 225.

38. Mottram, D.S.; Nobrega, I.C. In *Food Flavors: Formation, Analysis and Packaging Interactions.* E.T. Contis, C.T. Ho, C.J. Mussinan, T.H. Parliament, F. Shahidi, A.M. Spanier, eds. Elsevier Publ: Amsterdam, 1998; p. 483.

39. Farmer, L.J.; Patterson, R.L.S. *Food Chem* **1991**, 40, 201.

40. Rizzi, J.P.; Sanders, R.A. In *Flavour Science: Recent Developments.* A.J. Taylor; D.S. Mottram, Eds. Royal Society of Chemistry: London, 1996; p. 206.

41. Rizzi, J.P.; Bunke, P.R. In *Food Flavors: Formation, Analysis and Packaging Interactions.* E.T. Contis, C.T. Ho, C.J. Mussinan, T.H. Parliament, F. Shahidi, A.M. Spanier, Eds. Elsevier Publ: Amsterdam, 1998; p. 535.

42. Reil, G.; Berger, R.G. In *Flavour Science: Recent Developments.* A.J. Taylor; D.S. Mottram, eds. Royal Society of Chemistry: London, 1996; p. 97.

43. Demyttenaere, J.C.R.; Koninckx, I.E.I.; Meersman, A. In *Flavour Science: Recent Developments.* A.J. Taylor; D.S. Mottram, Eds. Royal Society of Chemistry: London, 1996; p. 105.

44. Benz, I.; Mulheim, A. In *Flavour Science: Recent Developments.* A.J. Taylor; D.S. Mottram, eds. Royal Society of Chemistry: London, 1996; p. 111.

45. Stam, H.; Boog, A.L.G.M.; Hoogland, M. In *Flavour Science: Recent Developments.* A.J. Taylor; D.S. Mottram, Eds. Royal Society of Chemistry: London, 1996; p. 122.

46. van der Schaft, P.H.; Goede, de H.; Burg, N. ter In *Flavour Science: Recent Developments.* A.J. Taylor; D.S. Mottram, Eds. Royal Society of Chemistry: London, 1996; p. 134.

FLAVOR FORMATION

Chapter 3

Lipids in Flavor Formation

Fereidoon Shahidi

Department of Biochemistry, Memorial University of Newfoundland,
St. John's, NF A1B 3X9, Canada

Lipids not only serve as a source of condensed energy, they also
provide essential fatty acids as well as mouthfeel and other
attributes to food. Flavor effect of lipids is related to the interaction
of food components with one another as well as lipid breakdown
products via their participation in Maillard reaction under high
temperature conditions experienced during processing. These
reactions occur during frying, grilling and other modes of heat
processing of foods. In addition, lipids may contribute to the flavor
of fresh foods via lipoxygenase-assisted oxidation. Lipoxygenases
are present in plants such as soybean and in fish gills, among
others; their interaction with cis,cis-1,4-pentadiene moieties of
lipids leads to the formation of stereospecific products.
Furthermore, breakdown of lipids under thermal or photooxidative
conditions produces an array of products, all of which are odor-
active and may contribute to off-flavor development in both raw and
processed foods. In particular, aldehydes and other carbonyl
compounds serve as indicators of flavor deterioration of many
foods. In addition, lipolytic reactions provide another route by
which flavor reversion of food lipids may occur, such as those in
butter and dairy products.

Lipids not only serve as a source of condensed energy, essential fatty acids and fat-
soluble vitamins, they also provide a reaction medium for interaction of different food
components and generation of aroma compounds. The influence of lipids on flavor
attributes of foods may be through aroma and aroma effects, flavor character, flavor
masking, as well as flavor release and development. Lipids also play a role during the
processing of foods and affect storage stability of products. Both polar and non-polar
lipids contain fatty acids with varying chain lengths and degrees of unsaturation, but
polar lipids generally contain a higher proportion of unsaturated fatty acids (1).
Participation of lipids in aroma generation involves different routes and mechanisms.

These include lipid oxidation which may be subdivided into autoxidation, thermal oxidation, photooxidation and lipoxygenease-derived oxidation (*2-5*). In this case, depending on the processing technique employed, the flavor effects might be quite different. For example, oxidation taking place in oils during storage is different from those of the oils during frying. In addition, lipids may be hydrolyzed/lipolyzed and produce free fatty acids, lactones and other compounds.

Upon heat processing, lipids and their breakdown products may participate in Maillard reaction, thus impacting the formation of different aroma compounds and the production of desirable flavors (*3, 6-9*). Although flavor of foods is due to cumulative effects of aroma, taste, texture and tactile, it is the aroma that is most readily perceived even prior to the consumption of food.

In this overview, an attempt is made to provide a concise account of the role of lipids in flavor formation in foods.

Autoxidation and Photooxidation Reactions and Flavor Impact

Autoxidation and photooxidation reactions are closely related to one another since the products formed are very similar, but not the same, because of different mechanisms involved. While autoxidation requires unsaturated fatty acids and triplet oxygen along with an initiator (e.g. heat, light, transition metal ions, etc.), photooxidation requires unsaturated fatty acids, singlet oxygen and a photosensitizer (e.g. chlorophyll, methylene blue, erythrosine, rose benegal, etc.) (*3*).

The autoxidation reaction involves a chain reaction mechanism in which a lipid free radical is formed during the initiation step followed by its interaction with triplet oxygen to afford a lipid hydroperoxide and a new lipid radical, known as propagation reaction. Finally, the radical species may combine with one another to produce non-radical products in order to terminate the process. The hydroperoxides formed in the reaction sequence are known as primary products of oxidation and are very unstable because of a weak oxygen-oxygen bond (*10-13*). These hydroperoxides subsequently degrade to an array of products such as hydrocarbons, aldehydes, alcohols, etc. (Figure 1). While the primary products of lipid oxidation are tasteless and odorless, the secondary oxidation products are aroma-active and their effect on flavor of foods depends on the threshold values of compounds involved (see Table I).

Food antioxidants and free radical quenchers donate a hydrogen atom to the lipid free radical and terminate the reaction; the antioxidant radicals formed are, however, stable and do not easily undergo further reaction. Thus, antioxidants might be used to control autoxidation of food lipids. The rate of autoxidation of fatty acids depends on their degree of unsaturation (Table II).

In the photooxidation or photooxygenation reaction a singlet oxygen is produced from the triplet oxygen by interaction of light and a photosensitizer such as chlorophyll (*3*). The photooxidation proceeds via an "ene" reaction in which the singlet oxygen adds to an olefinic carbon atom with subsequent integration of the double bond and change from *cis* to *trans* configuration (Figure 2). This reaction can only be terminated by the presence of a singlet oxygen quencher such as β-carotene, but it is unaffected by the usual phenolic antioxidants. The photooxidation reaction is much faster than autoxidation and the degree of unsaturation of lipids plays a minor role in the speed at which the reaction proceeds (see Table II).

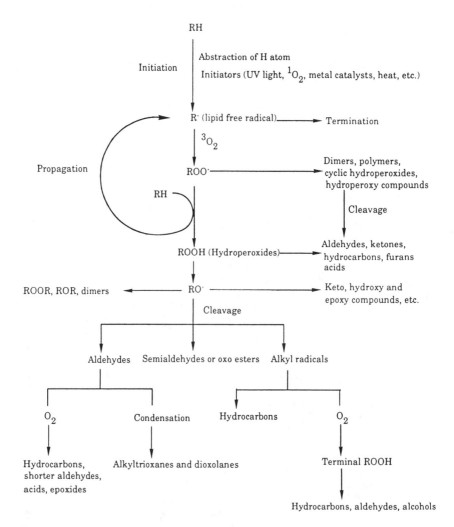

Figure 1. The mechanism of lipid oxidation and formation of primary and secondary degradation products.

$$Sen + h\nu \longrightarrow Sen^* + {}^3O_2 \longrightarrow Sen + {}^1O_2$$

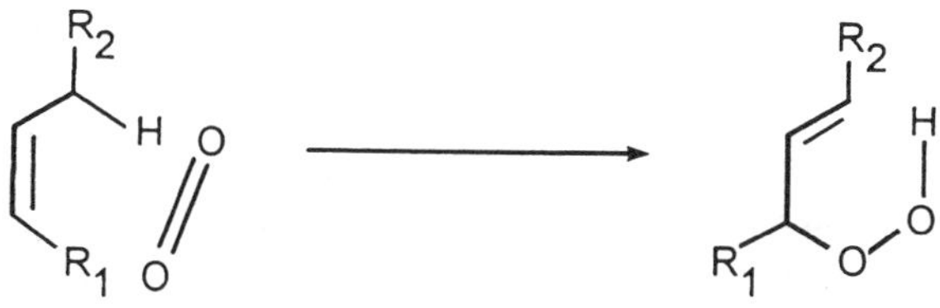

Figure 2. The mechanism of photoxidation and conversion of cis to trans configuration. This reaction is unaffected by antioxidants, but is inhibited by quenchers of singlet oxygen such as β-carotene.

Table I. Threshold values of selected classes of oxidation
products of lipids

Compound	Threshold, ppm
Hydrocarbons	90-2150
Furans	2-27
Vinyl alcohols	0.5-3
Alkenes	0.02-9
2-Alkenals	0.04-1.0
(E,E)-2,4-Alkadienals	0.04-0.3
Alkadienal, isolated	0.002-0.3
(Z)-Alkenals, isolated	0.0003-0.1
(E,Z)-Alkadienals	0.002-0.006
Vinyl ketones	0.00002-0.007

Table II. Relative rates of oxidation by triplet oxygen (autoxidation) and
singlet oxygen (photooxidation)

Oxygen	C18:1	C18:2	C18:3
Triplet	1	27	77
Singlet	3×10^4	4×10^4	7×10^4

Table III. Monohydroperoxides formed by the reaction of triplet oxygen
($^3O_2$) and singlet oxygen ($^1O_2$) with unsaturated fatty acids

Fatty Acid	Position		Proportion, %	
	HOO-group	Double bond	$^3O_2$	$^1O_2$
Oleic acid	8	9	27	
	9	10	23	48
	10	8	27	52
	11	9	27	
Linoleic acid	8	9, 12	1.5	
	9	10, 12	46.5	32
	10	8, 12	0.5	17
	12	9, 13	0.5	17
	13	9, 11	49.5	34
	14	9, 12	1.5	
Linolenic acid	9	10, 12, 15	31	23
	10	8, 12, 15		13
	12	9, 13, 15	11	12
	13	9, 11, 15	12	14
	15	9, 12, 16		13
	16	9, 12, 14	46	25

Figures 3-5 compare the autoxidation and photooxidation reactions leading to peroxide formation for oleic, linoleic and linolenic acids, respectively (e.g., 3). The number and concentration of hydroperoxides so produced depend on the nature of the fatty acids involved (see Table III). Obviously, breakdown of these hydroperoxides leads to the formation of different carbonyl compounds, some of which are listed in Table IV.

Both hydroperoxides (primary products) and their breakdown products (secondary products) affect the safety and wholesomeness of foods as they also react with DNA, carbohydrates and proteins and cause mutation and other deleterious reactions. They may also exert a toxic and carcinogenic effect. In addition, oxidation of lipids leads to the formation of off flavors, loss of essential fatty acids and fat-soluble vitamins (2).

Assays to follow lipid oxidation in foods include quantitation of changes in the starting materials, namely composition of fatty acids, iodine volume, weight gain and oxygen consumption. In addition, the primary products of lipid oxidation may be quantitated by monitoring the peroxide value, using potassium iodide and subsequent titration with a standardized sodium bisulfite. Individual hydroperoxides may also be determined using high performance liquid chromatography followed by possible post-column derivitization and detection. In addition, conjugated dienes and trienes may be determined by monitoring absorbance values at 234 and 268 nm, respectively (15).

Determination and quantitation of secondary products of autoxidation might be followed using the 2-thiobarbituric acid and p-anisidine tests, or reaction with 2,4-dinitrophenylhydrazine or reaction with N-N-dimethyl phenylenediamine (DDP) (16) or determination of individual carbonyl compounds. In general, pentane and hexanal are used for determination of the extent of oxidation of omega-6 fatty acids while propanal is used to assay the breakdown of omega-3 fatty acids (17).

Finally, the overall oxidation may be monitored using TOTOX value (2 PV + p-AnV), infrared (IR) spectroscopy or ^{1}H NMR spectroscopy, as well as the Ranciment or OSI determinations of the induction periods of oxidation. Of these, the industry is mostly using TOTOX value and OSI or Rancimat machines in order to monitor oxidation (18).

Hydrolytic Rancidity

Lipids, under different conditions, especially in the presence of enzymes and adequate moisture or without enzyme may be hydrolyzed to produce free fatty acids, ketones or lactones, depending on the nature of the fatty acids involved. These breakdown products impact the flavor of different lipids, particularly dairy products where short-chain fatty acids are a major contributor (see Figure 6). Thus, production of butyric acid from lipolysis of butter leads to the formation of a sharp rancid note (19).

In dry-cured-ham, formation of free fatty acids and related compounds might be responsible for development of a desirable flavor in the products. These compounds are formed in such products via fermentation reactions, as explained by Toldra et al. (20).

During deep-fat frying many changes occur in the physical and chemical characteristics of the oil and the food that is being processed. These changes include

30

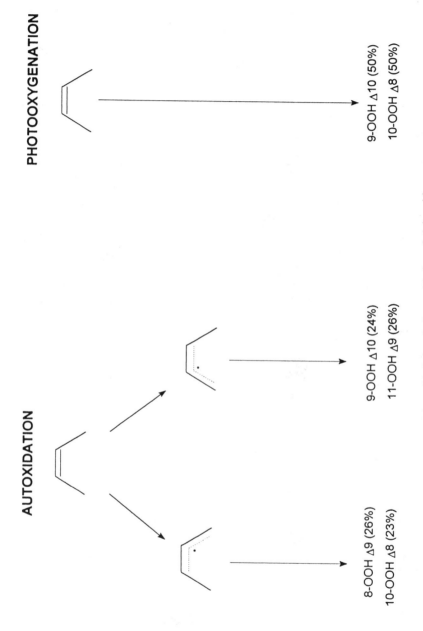

Figure 3. Autoxidation and photooxidation of oleic acid.

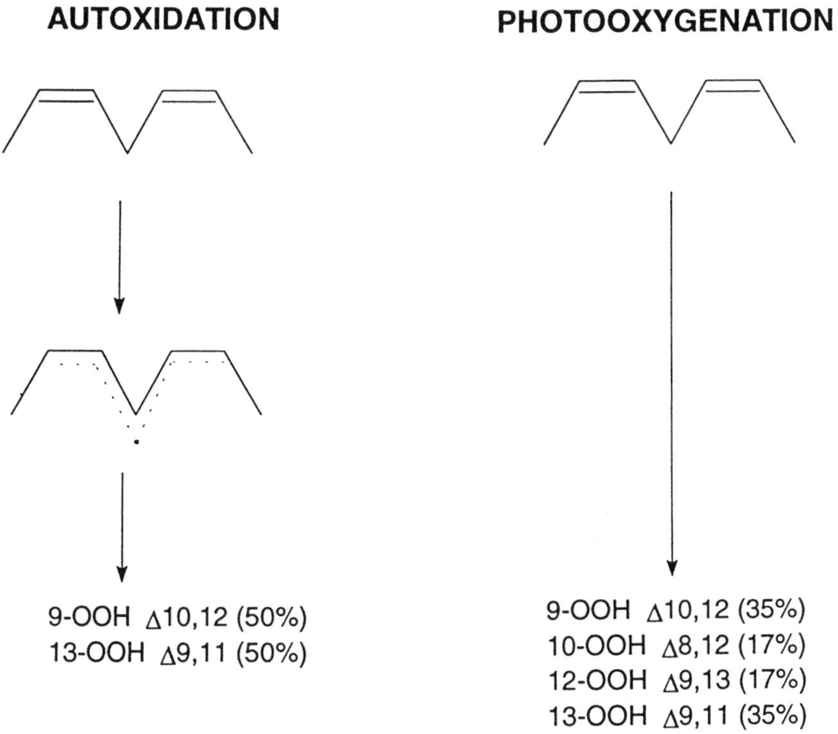

Figure 4. Autoxidation and photooxidation of linoleic acid.

32

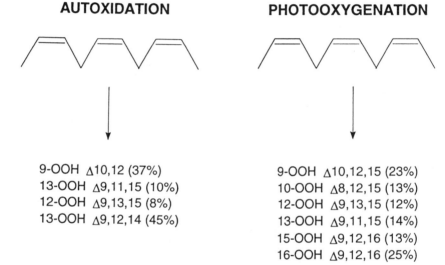

Figure 5. Autoxidation and photooxidation of linolenic acid.

33

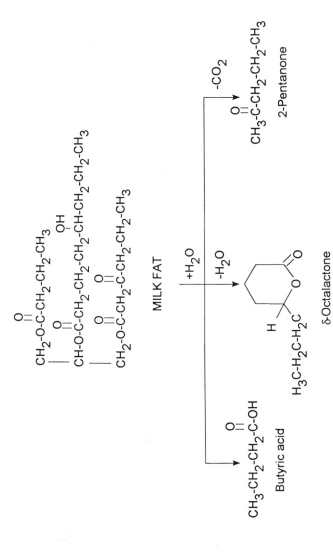

Figure 6. Formation of impact volatile flavor compounds from milkfat generated via hydrolytic cleavage of acylglycerols.

Table IV. Selected aldehydes and substituted furans from breakdown of individual hydroperoxides of oleate, linoleate and linolenate

Oleate		Linoleate		Linolenate	
Compound	Source	Compound	Source	Compound	Source
Heptanal	?	Ethanal	?	Ethanal	?
Octanal	11-OOH	Pentanal	13-OOH	Propanal/Acrolein	15/16-OOH
Nonanal	9/10-OOH	Hexanal	12/13-OOH	Butanal	?
2-Nonenal	?	2-Heptenal	12-OOH	2-Butenal	15-OOH
Decanal	8-OOH	2-Pentylfuran	?	2-Pentenal	13-OOH
2-Decenal	9-OOH	2-Octenal	?	2/3-Hexenal	12/13-OOH
2-Undecenal	8-OOH	2-Nonenal	9/10-OOH	2-Butylfuran	?
		2,4-Nonadienal	?	2,4-Heptadienal	12-OOH
		2,4-Decadienal	9-PPH	3,6-Nonadienal	9/10-OOH
				Decatrienal	9-OOH

oxidation, absorption, dehydration, polymerization as well as hydrolysis and volatilization. Hydrolysis of lipids during deep-fat frying produces free fatty acids, monoacylglycerols, diacylglycerols and glycerol. Some products of hydrolysis, such as glycerol, may vaporize and others may undergo further reactions, including autoxidation. Since glycerol volatilizes at above 150°C, the reaction equilibrium is shifted in favor of other hydrolysis products. Therefore, hydrolysis occurring during deep-fat frying has a considerable effect on flavor quality of both the lipids and the food (2).

Conversion of Certain Lipid Oxidation Products to Other Compounds

Lipids containing unsaturated fatty acids may oxidize and produce an array of products. These products are often unsaturated themselves and as such may undergo further degradation. As an example, 2,4-decadienal may be further oxidized to afford 2-octenal which would in turn produce hexanal (Figure 7) which may in turn be oxidized to hexanoic acid and gamma-hexalactone. Formation of 2-butene-1,4-dial from 2,4-decadienal, along with hexanal from 2,4-decadienal may also be contemplated (9, 21-22). Furthermore, 2-octenal may undergo further oxidation which eventually leads to the formation of heptanal.

Lipoxygenase-Assisted Oxidation of Food Lipids

Lipoxygenases are found abundantly in different species of plant (23-25), animal (26) and fish (27-29). Table V summarizes the sources of lipoxygenase in different species. Individual lipoxygenases may have some specificity in the formation of individual hydroperoxides (see Table VI). These lipoxygenases affect the cis, cis-1,4-pentadiene moiety of fatty acids and generate myriad of aroma active compounds with defined stereospecific nature.

Table V. Selected sources of plant and animal lipoxygenase

Plant	Mammal	Fish
Alfalfa	Leucocytes	Bass
Beans	Lungs	Blue Gill
Cucumber	Platelets	Catfish
Grape	Reticulocytes	Emerald Shiner
Mango	Skins	Perch
Mushroom		Salmon
Peas		Trout
Potato		
Soybean		
Tobacco		
Tomato		

36

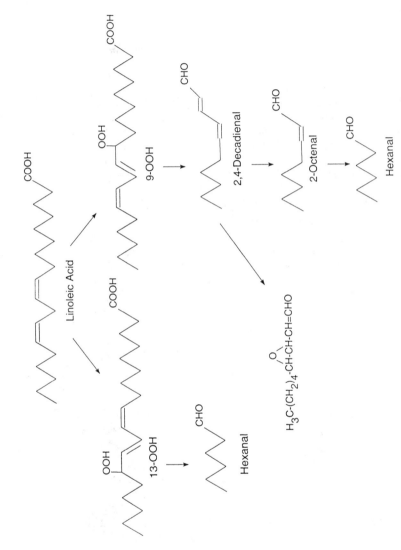

Figure 7. Oxidation of linoleic acid and formation of 2-octenal and hexanal from 2,4-decadienal.

Table VI. Occurrence and Properties of Various Lipoxygenases

Food	pH optimum	Peroxidation specificity %[a]	
		9-LOOH	13-LOOH
Soybean, L-1	9.0	5	95
Soybean, L-2	6.5	50	50
Peas, L-2	6.5	50	50
Peanut	6.0	0	100
Potato	5.5	95	5
Tomato	5.5	95	5
Wheat	6.0	90	10
Cucumber	5.5	75	25
Apple	6.0	10	90
Strawberry	6.5	23	77
Gooseberry	6.5	45	55

Lipoxygenases have several effects on foods, both desirable and undesirable. The desirable function of lipoxygenase relates to its ability to bleach wheat and soybean flours and to assist in the formation of disulfide bonds in gluten during dough formation. The undesirable actions of lipoxygenase in food relate to the destruction of chlorophyll and carotene, development of oxidative flavors and aromas, oxidative damage to proteins and vitamins and oxidation and destruction of essential fatty acids.

The specificity of products formed from lipoxygenase-assisted oxidation of linolenic acid may be demonstrated in fruits and their cut tissues. Production of (E)-2-hexenal in fresh tomatoes and (E,Z)-2,6-nonadienal in cucumbers by site-specific hydroperoxidation is dictated by a lipoxygenase and a subsequent lyase cleavage reaction (Figure 8). Disruption of tissues and initial generation of carbonyl compounds may be followed by other reactions, and since successive reactions occur, overall aroma of the material may change with time. As an example, (E,Z)-2,6-nonadienal may be converted to its corresponding alcohol via the action of an alcohol dehydrogenase (Figure 8). The alcohol so formed has a higher detection threshold and heavier aromas than the precursor carbonyl compound.

In animal tissues, especially fish, lipoxygenase may produce a fishy aroma via reaction with omega-3 fatty acids. The characteristic aromas are generally due to 2,4,7-decatrienal isomers and (Z)-4-heptenal may potentiate the fishy character of the former compound. It should also be noted that the very fresh seafood aroma is due to a group of enzymatically-derived aldehydes, ketones and alcohols and these are very similar to the C_6, C_8 and C_9 compounds produced via the action of plant lipoxygenases (30). Thus, flavor attributes such as melony, heavy plant like, and fresh fish aromas are often perceived by evaluators for freshly harvested fish. Lipoxygenase found in fish may first produce alcohols that are then converted to the corresponding carbonyl compounds such as cis-1,5-Octadien-3-one. A proposed mechanism for the biogenesis of some fresh seafood aroma compounds from eicosapentaenoic acid (EPA) is given in Figure 9 (29, 31).

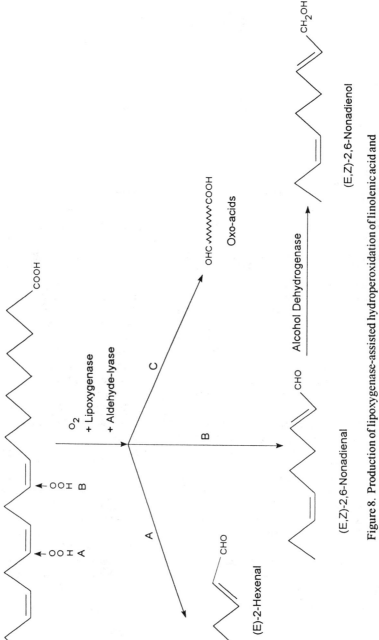

Figure 8. Production of lipoxygenase-assisted hydroperoxidation of linolenic acid and their subsequent cleavage to produce specific aldehydes. Conversion of the aldehyde to alcohol with subtle flavor change is shown.

39

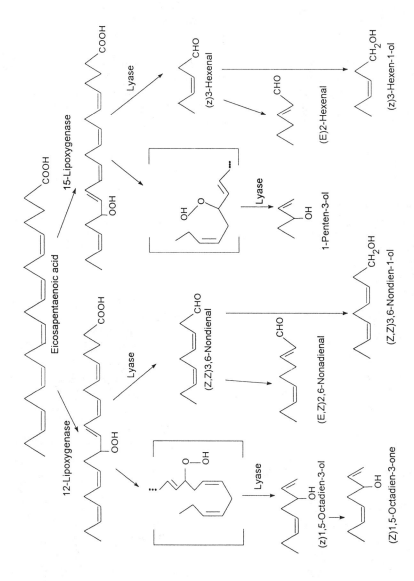

Figure 9. Formation of fresh seafood aroma via enzymatic breakdown of eicosapentaenoic acid.

Role of Lipids and Their Degradation Products in Maillard Reaction

The reaction of free amino acids, amines, peptides and proteins with reducing sugars leads to the formation of Maillard reaction products (*32*). Generally, free amino acids produce aldehydes, hydrogen sulfide and ammonia which may subsequently participate in the Maillard reaction.

The role of lipids in Maillard reactions was demonstrated by Mottram and co-workers (e.g., *33, 34*) where they heat processed meat, as such, or first defatted it with hexane (to remove only the triacylglycerols, but retaining the phospholipids) or methanol-chloroform (to remove both neutral and polar lipids) prior to cooking. The sensory impact of the processed materials was quite different. The control sample, upon heat processing, gave a meaty aroma, while the intensity of the meaty aroma was reduced in the hexane-extracted counterpart. Meanwhile, the methanol-chloroform extracted sample had very little meaty note, but possessed a sharp roast and biscuit-like odor. In particular, the concentration of dimethypyrazine in the headspace volatiles was significantly increased with a concurrent decrease in the content of lipid oxidation products. Thus, it was concluded that phospholipids were primarily responsible for the development of a meaty aroma.

To confirm these effects, Farmer and Mottram (*4*) used a model system in which ribose was reacted with cysteine, as such, or in the presence of beef triacylglycerols (BTAG) or beef phospholipids (BPL). There was a marked reduction in the amount of thiols when phospholipids, and to a lesser extent tricylglycerols, were present (Table VII). The compounds that were formed only in the presence of lipids were 2-pentylpyridine, 2-alkylthiophenes, alkenylthiophenes, pentylthiapyran and alkanethiols. Furthermore, the impact of BPL was much greater than that of BTAG in affecting the flavor of systems under investigation. The reactions of lipids and lipid breakdown products with Maillard precursors may be due to the participation of carbonyl compounds. These compounds may participate as such or may degrade further to other products prior to their involvement in the reaction.

Table VII. Relative concentration of selected acyclic and heterocyclic volatile products from the reaction of cysteine with ribose alone or in the presence of beef triacylglycerols (BTAG) or beef phospholipids (BPL)

Compound	No Lipid	BTAG	BPL
2-Pentanone, 3-mercapto	1	0.77	0.47
3-Pentanone, 2-mercapto	1	0.72	0.49
2-Furylmethanethiol	1	0.67	0.63
3-Furanethiol, 2-methyl	1	0.40	0.15
2-Thiophenethol	1	0.32	0.03
2-Pentylpyridine	0	0.09	1
2-Pentylthiophene	0	0	1
2-Hexylthiophene	0	0.15	1
2-Pentyl-2-H-thiapyran	0	0.10	1

Table VIII. Products of interaction of 2,4-decadienal with cysteine

Compound	Amount generated (mg/mol)
FURANS	
2-Butyl	12.8
2-Pentyl	6.4
2-Hexyl	t
THIOPHENES	
Unsubstituted	3.5
Tetrahydro, 3-one	10.5
2-Butyl	57.2
2-Formyl-3-methyl*	29.8
2-Pentyl	13.1
2-Hexyl	42.0
2-Heptyl	1.8
2-Formyl-5(or 3)-pentyl	15.6
THIAZOLES	
Unsubstituted	25.6
2-Acetyl	2.2
3-Methyl,iso	2.0
CYCLIC POLYSULFIDES	
3,5-Dimethy-1,3,4-trithiolane isomers	141.0
3-Methyl-5-pentyl-1,2,4-trithiolane	14.3
2,4,6-Trimethylperhydro-1,3,5-thiadizine	828.5
2,4,6-Trimethylperhydro-1,3,5-dithiazine	284.2
2,4-Dimethyl-6-perlyeperhydro-1,3,5-dithiazine	18.9
2-Pentyl-4,6-dimethylperhydro-1,3,5-dithiazine	28.7
PYRIDINES	
2-Pentyl	501.5

[a]Adapted from Zheng and Ho, 1989.

In another study, Zhang and Ho (*21*) reacted cysteine directly with 2,4-decadienal. A significant amount of 2-substituted heterocyclic compounds was formed (Table VIII). Thus, participation of lipid degradation products is Maillard reaction is well established.

The mechanism by which alkyl substituted heterocyclics are formed is varied and depends on the compounds involved. Formation of several 2-alkyl-substituted heterocyclic compounds from the reaction of decadienal with NH_3 and H_2S which were formed from Strecker degradation or other sources has been documented (*4*). Zhang and Ho (*21*) have also shown that alkylpyridins are formed from the interaction of amino acids with decadienal. Formation of alkylpyrazine via the involvement of alkanals has been documented (*35*). Similarly, acetaldehyde may react with 2,4-

42

decadienal to form 2-pentylbenzaldehyde (*22*). Participation of other lipid- derived products in Maillard reactions has also been shown (*21, 22, 35*). Furthermore, Maillard reaction products so formed may act as important antioxidants in order to stabilize lipids (*36, 37*).

Conclusions

Lipids play an important role in flavor formation. Both desirable and undesirable aroma-active compounds are formed. Based on our current knowledge and mechanistic views, it is possible to use lipids to modify flavor of foods and it is feasible to control oxidation, where and when it might be necessary, in order to prevent the formation of off-flavor compounds. Use of lipids for biogeneration of aroma, using selected enzymes, is of considerable importance for the food and flavor industries.

References

1.   Shahidi, F.; Shukla, N.K.S. *Inform* **1996**, *7*, 1227-1231.
2.   Chow, C.K.; Gupta, M.K. In *Technological Advances in Improved and Alternative Sources of Lipids*; Kamel, B.S. and Kakuda, Y, Eds.; Chapman and Hall; London, 1994; pp 329-359.
3.   Gunstone, F.D. *J. Am. Oil Chem. Soc.* **1984**, *62*, 441-447.
4.   Gardner, H.W. *J. Am. Oil Chem. Soc.* **1996**, *73*, 1347-1357.
5.   Hsieh, R.J. In *Lipids in Food Flavors*; Ho, C-T. and Hartman, T.G., Eds.; ACS Symposium Series 558; American Chemical Society: Washington, DC, 1994; pp 30-48.
6.   King, D.L.; Hahm, T.S.; Min, D.B. In *Shelf-Life Studies of Foods and Beverages: Chemical, Biological, Physical and Nutritional Aspects*; Charalambous, G., Ed.; Elsevier: Amesterdam; pp 629-705.
7.   Farmer, L.J.; Mottram, D.S. *J. Sci. Food Agric.* **1990**, *53*, 505-525.
8.   Farmer, L.J.; Mottram, D.S. In *Trends in Flavour Research;* Maarse, H. and van der Heij, D.G., Eds.; Elsevier: Amsterdam, 1994; pp 313-326.
9.   Zhang, Y.; Ritter, W.J.; Barker, C.C.; Traci, P.A.; Ho, C-T. In *Lipids in Food Flavors;* Ho, C-T. and Hartman, T.G., Eds.; ACS Symposium Series 558; American Chemical Society: Washington, DC, 1994; pp 49-60.
10.   Chan, H.W-S.; Coxon, D.T. In *Autoxidation of Unsaturated Lipids;* Chan, H. W-S., Ed.; Academic Press: London, 1987; pp 17-50.
11.   Frankel, E.N. *J. Am. Oil Chem. Soc.* **1984**, *61*, 1908-1917.
12.   Frankel, E.N. In *Flavor Chemistry of Fats and Oils;* Min, D.B. and Smouse, T.H., Eds.; American Oil Chemists' Society: Champaign, IL, 1985; pp 1-37.
13.   Frankel, E.N.; Neff, W.E.; Selke, E. *Lipids* **1981**, *16*, 279-285.
14.   Ho, C-T.; Chen, Q. In *Lipids in Food Flavors;* Ho, C-T. and Hartman, T.G., Eds.; ACS Symposium Series 558; American Chemical Society: Washington, DC, 1994; pp 2-14.
15.   Frankel, E.N. *Trends in Food Science* **1993**, *4*, 220-225.
16.   Miyashita, K.; Kanda, K.; Takagi, T. *J. Am. Oil Chem. Soc.* **1991**, 748-751.
17.   Frankel, E.N.. *J. Am. Oil Chem. Soc.* **1993**, *70(8)*, 767-772.

18. Shahidi, F.; Wanasundara, U.N. In *Natural Antioxidants: Chemistry and Health Effects, and Applications;* Shahidi, F., Ed.; AOCS Press: Champaign, IL, 1997, pp 379-396.
19. La Grange, W.S.; Hammond, E.G. In *Shelf Life Studies of Foods and Beverages, Chemical, Biological, Physical and Nutritional Aspects;* Charalambous, G., Ed.; Elsevier: Amsterdam, 1995; pp 1-20.
20. Toldra, F.; Flores, M.; Sanz, Y. *Food Chem.* **1997,** *59,* 523-530.
21. Zhang, Y.; Ho, C-T. 1989. *J. Agric. Food Chem.* **1989,** *37,* 1016-1020.
22. Yu, T-H.; Lee, M-H.; Wu, C-M.; Ho, C-T. In *Lipids in Food Flavors*; Ho, C-T. and Hartman, T.G., Eds.; ACS Symposium Series 558; American Chemical Society: Washington, DC, 1994; pp 61-76.
23. Yoon, S.; Klein, B.P. *J. Agric. Food Chem.* **1979,** *27,* 955-962.
24. Vernnooy-Gerritsen, M.; Bos, A.L.M.; Veldink, G.A.; Vliegenthart, J.F.G. *Plant Physiol.* **1983,** *73,* 262-267.
25. Vernooy-Gerritsen, M.; Leunissen, J.L.M.; Veldink, G.A.; Vliegenthart, J.F.G. *Plant Physiol.* **1984,** *76,* 1070-1079.
26. Whitaker, J.R. In *Oxidative Enzymes in Foods;* Robinson, D.S. and Eskin, N.A.M., Eds.; Elsevier Science Publisher, Ltd.: London, 1991; pp 175-215.
27. German, J.B.; Kinsella, J.E. 1985. *J. Agric. Food Chem.* **1985,** *33,* 680-683.
28. German, J.B.; Creveling, R.K. *J. Agric. Food Chem.* **1990,** *54,* 1889-1891.
29. Josephson, D.B.; Lindsay, R.C. In *Biogeneration of Aromas*; Parliament, T.H. and Croteau, R., Eds.; ACS Symposium Series 317; American Chemical Society: Washington, DC, 1986; pp 200-217.
30. Lindsay, R.C. In *Food Chemistry,* 3rd Edition; Fennema, O.R., Ed.; Marcel Dekker: New York, 1996; pp 723-765.
31. Durnford, E.; Shahidi, F. In *Flavor of Meat, Meat Products and Seafoods*; Shahidi, F., Ed.; Blackie Academic and Professional: London, 1998; pp 131-158.
32. Maillard, L.C. *Compt. Rend.* **1912,** *154,* 66-68.
33. Mottram, D.S.; Edwards, R.A. *J. Sci. Food Agric.* **1983,** *34,* 517-522.
34. Mottram, D.S. In *Volatile Compounds in Foods and Beverages*; Maarse, H., Ed.; Marcel Dekker: New York, 1991; pp 107-177.
35. Ho, C.-T.; Bruechert, L.J.; Zhang, Y.; Chiu, E.-M. In *Thermal Generation of Aromas*; Parliament, T.M.; McGorrin, R.J.; and Ho, C.-T., Eds.; ACS Symposium Series 407; American Chemical Society: Washington, DC, 1989; pp 105-113.
36. Bailey, M.E.; Clarke, A.D.; Kim, Y.S.; Fernando, L. *Natural Antioxidants: Chemistry, Health Effects, and Application*; Shahidi, F., Ed.; AOCS Press: Champaign, IL, 1997; pp 296-310.
37. Bailey, M.E. In *Flavor of Meat, Meat Products and Seafoods;* Shahidi, F., Ed.; Blackie Academic and Professional: London, 1998; pp 267-289.

Chapter 4

Critical Flavor Compounds in Dairy Products

T. H. Parliment[1] and R. J. McGorrin[2,3]

[1]Technology Center, Kraft Foods, Inc.,
555 South Broadway, Tarrytown, NY 10591
[2]Technology Center, Kraft Foods, Inc., 801 Waukegan Road, Glenview, IL 60025

The aroma composition of dairy products has been studied since the mid-1950's. In the ensuing years, numerous volatile compounds of milk, cultured dairy products and cheese have been identified and published in the technical literature. The great majority of these works have discussed the qualitative and quantitative composition, but not the flavor relevance of the identifications. The purpose of this chapter is to review the recent published technical literature with the goal of providing a review of the critical flavor compounds of cow's milk, cream, buttermilk, yogurt and various cheeses. Particular emphasis is placed on studies involving organoleptic–gas chromatographic assays such as AEDA, CharmAnalysis, OAV and flavor dilution factor studies.

A significant number of publications report volatile compounds which have been identified in milk and dairy products (*1-6*). However, while the flavor chemistry and sensory impact of compounds identified in dairy products has been well documented in three recent reviews (*7-9*), little data is available in a single source regarding flavor components in these materials that have sensory relevance. We undertook this review to assist the flavor researcher and creative flavorist in identifying the critical volatile flavor components in dairy products, and to put into context the volatile compound information which has previously been reported in the literature.

[3]Corresponding author.

Sample Preparation Techniques

Volatile isolation methods such as dynamic headspace concentration and simultaneous distillation-extraction (SDE) have been commonly used for the analysis of flavors (*10,11*). Volatile headspace trapping on Tenax adsorbent resins has become a mainstay for flavor analysis laboratories because it is moderately sensitive, can be performed rapidly, and is a technique which is not prone to generate thermal artifacts, as frequently happens with dairy products using SDE. Because flavor researchers are interested in quantifying the most potent aroma components, many have selected headspace trapping to compare flavor differences among dairy products. Alternatively, dairy volatiles can be isolated using high-vacuum distillation and condensation at liquid nitrogen temperature. In any event, care must be taken to avoid thermal artifact formation during isolation of delicate dairy flavors. Since these components are present at low concentrations, prolonged heating of dairy samples during isolation to increase flavor volatility and recovery can potentially alter their composition. Dairy flavor researchers have used simultaneous distillation under vacuum with continuous sample feeding to minimize thermal artifact formation (*12*).

Sensory-Directed Analytical Techniques

Following sample concentration, flavor chemists typically rely on gas chromatography-mass spectrometry (GC/MS) to simultaneously separate and identify new components in dairy flavors. However due to the complexity of these dairy isolates, some flavor researchers have relied on sensory-directed techniques to identify sensory-significant regions of the chromatogram, prior to time-consuming mass spectral interpretation and structural elucidation. This sensory-directed analytical approach is outlined in Figure 1.

Gas Chromatography-Olfactometry (GCO)

An initial approach used to distinguish odor-significant compounds from the less important volatiles present in dairy flavor extracts is the application of gas chromatography/olfactometry (GCO) (*10,13*). Simultaneous with the high resolution gas chromatography separation of a volatile extract, the odor of individual compounds is assessed by sniffing the effluent of the GC column in parallel with electronic detection. This technique enables the detection of odor-active volatiles, the determination of their odor qualities, and a correlation of sensory data with retention indices and GC-MS identifications.

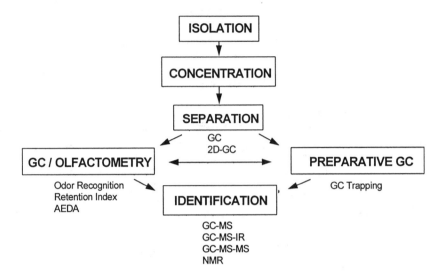

Figure 1. Sensory-directed flavor analysis strategy

Odor Dilution Techniques: CharmAnalysis and AEDA

Since it is difficult to evaluate the intensities of individual odorants during sniffing, odor dilution techniques have been further developed. Combined Hedonic Aroma Response Measurements (CharmAnalysis) developed by Acree (*14*) and Aroma Extract Dilution Analysis (AEDA) developed by Grosch (*15,16*) enable the relative odor importance of individual odorants to be determined without knowledge of their chemical structures. The techniques use serial dilutions of the volatile fraction, followed by GCO evaluation after each successive dilution. The process is repeated until no detectable odor is perceivable in the GC effluent. The highest dilution at which a compound can be smelled is defined as its flavor dilution (FD) factor in the AEDA method. While the duration of the odor is also factored into the CharmAnalysis, the peak maximum of a CharmAnalysis value is identical to the FD factor. Since both of these techniques rank the odorants present in a dairy extract according to their relative flavor potencies, GC-MS analyses can become focused to identification of only the key odorants exhibiting high FD factors or Charm values.

Odor Activity Values (OAV)

An approach that further augments the AEDA odor dilution technique was developed by Grosch and co-workers (*17*), which determines the odor activity values

(OAVs) of the key odorants. This technique requires access to published threshold values for specific compounds, or for newly identified compounds, the determination of individual odor thresholds in water. Many of the aroma thresholds for volatiles identified in dairy products have been previously reported (3,18).

The odor activity value is defined as the ratio of a flavor compound's concentration to its odor threshold:

$$OAV = \frac{Compound\ Concentration}{Odor\ Threshold}$$

The OAV is equivalent to terms such as aroma value, odor unit, or odor value as used by other flavor researchers (13,14).

The logarithm of the odor threshold (log OAV) is calculated to represent changes in concentration which are significant for olfactory discrimination. Odor activity follows a sigmoidal dose-response curve in that significant aroma responses require order-of-magnitude changes in concentration. Consequently, logarithmic functions more significantly represent meaningful sensory differences. Log OAV values >1 are indicative of compounds present at a concentration that greatly exceeds their thresholds, and therefore are likely to contribute significant flavor impact.

A critical prerequisite for calculating OAVs is accurate quantitative GC data. An internal standard having a chemical structure similar to the flavorant is used when working at concentration ranges of 0.1 ppm or higher. However, quantifying trace levels of volatile compounds requires standards prepared with stable isotopes (typically deuterium). Stable isotope dilution assays are analyzed by gas chromatography–selected ion mass spectrometry (SIM) in the chemical ionization mode.

Alternative Aroma Intensity Techniques: NIF and SNIF

Because odor dilution techniques are somewhat time-consuming, require repetitive analyses, and sometimes encounter reproducibility problems, another approach was recently developed by Pollien et al (19) to address these issues. The new methods are Nasal Impact Frequency (NIF) and Surface of Nasal Impact Frequency (SNIF). In contrast to odor dilution techniques, only one dilution level is used in NIF, however the GC-olfactometry experiment is replicated many times using eight or more evaluators, and the resulting individual aromagram values are then averaged to minimize reproducibility errors. Shorter time duration segments of the aromagram are performed in a single setting to eliminate sensory fatigue. NIF values are expressed as a normalized distribution (%), reflecting relative sensory impact. SNIF values combine sensory input with concentration as measured by GC peak abundance.

Critical Flavors in Milk and Cream

The underlying flavor of dairy products arises principally from the native volatile constituents in milk, and their degradation during processing and storage. Typical flavor volatiles in milk originate from the cow's metabolism with some influence from diet, such as from feed, grazing, or silage. The sensory profiles of various dairy products depend largely on the relative balance of flavor components derived from fat, protein or carbohydrate. Much of the production of dairy products involves the controlled fermentation of milk. Flavorants can be produced via four separate pathways (5):

- The first category include flavor compounds which are derived from lactose or citrate as a result of lactic flora (*lactococci* and *lactobacilli* bacteria), which generate compounds such as lactic acid, acetaldehyde, and diacetyl.
- The second group are flavor constituents which arise from unsaturated fatty acids in milk fat, via lipolysis or oxidation, to produce hydroxy acids, lactones, aldehydes and methyl ketones. Milkfat glycerides contain considerable quantities of C_4, C_6, C_8, and C_{10} fatty acids which are also strong flavor contributors to cultured products.
- A third pathway results from breakdown of casein, other milk proteins and amino acids, to provide Strecker aldehydes (*iso*-valeraldehyde, phenylacetaldehyde) and sulfur compounds derived from cysteine and methionine (methane thiol, dimethyl disulfide, methional).
- Finally, a fourth group of dairy flavorants arises from breakdown of sugars such as lactose, to yield compounds such as furanones, maltol and 2-acetylfuran.

Milk Flavor

Consumer acceptance of fresh milk is driven by its bland, yet characteristic flavor. As an ingredient, milk and its various derivatives are used as a base for the manufacture of cultured dairy products. The flavor of milk has been comprehensively reviewed by Badings (2,3) and Nursten (8). The key attributes of milk flavor are influenced by its existence as an emulsion of fat globules in a colloidal aqueous protein solution, accompanied by a slight sweet and salty taste from lactose and salts. Fresh milk is a well-blended composition of aromatic compounds which are present in concentrations at or below their flavor thresholds. Consequently, the flavor balance of fresh milk can be easily disrupted due to flavor compound defects produced by heat, light, oxygen, enzymatic and microbial activity,

Table I. Thermal Flavors Generated in Heat-Treated Bovine Milk

Thermal treatment		Temperature (°C)	Time (sec.)	Flavor Profile	Key Heat Flavor Components
Low-pasteurized (LP)		72	15	Bland, typical	H_2S (trace)
Medium-pasteurized (MP)		75	20	Slight cooked	H_2S
High-pasteurized (HP)		85	20	Distinct cooked, ketone, trace caramelized	C_9-C_{11} alkanals, butanone, 2,3-pentanedione
Ultra-high temp. (UHT; HTST)		142	4	Trace cooked	C_7, C_9 2-alkanones, dimethyl sulfide
Sterilized	initial	145	30	Strong caramelized, ketone-like	C_5,C_7,C_9,C_{11} 2-alkanones, pyrazines, furans, furanones, o-aminoacetophenone
	final	115	20 min.		

SOURCE: Adapted from ref. 3.

or contributed from feed sources. Souring or rancidity in milk develops when fatty acids are produced by the activity of native lipases on the lipid fraction.

Fresh milk flavor is described as delicate and well-blended. However, since milk as a beverage is consumed after pasteurization, heat-generated compounds are known to be formed which alter its flavor (20). Depending on how milk is pasteurized and thermally processed, different sensory profiles can be produced. Cooked off-flavors and flavor defects are generated in milk which has undergone various types of heat treatment. For example, low-pasteurized (LP) milk exhibits a very bland flavor, whereas high-pasteurized (HP) milk has a distinct cooked flavor accompanied by a slight ketone-like flavor. Ultra-high temperature processing yields a strong sulfurous aromatic, accompanied by a slightly sweet taste. Examples of flavors formed in heat-treated milk are summarized in Table I (3).

Previous research studies on milk flavor teach us that methyl ketones, aldehydes, delta-lactones, and dimethyl sulfide are important contributors to milk flavor (2). However, Moio et al (21) have recently applied CharmAnalysis to distinguish 14 odor-active volatiles in freshly secreted bovine, ovine, caprine and water buffalo milk. Of these, six aroma ccompounds were identified as important in bovine milk flavor (22) (Table II).

Table II. Potent Odorants in Fresh Bovine Milk

Mean* Charm Value	Compound	Odor Descriptor
1026	Ethyl hexanoate	Fruity, fragrant, sweet
1003	Ethyl butyrate	Vegetative, unripe fruit
379	Dimethylsulfone	Sulfurous, hot milk, burnt
122	Nonanal	Green, grass-like, tallowy
79	1-Octen-3-ol	Mushroom-like, musty, earth
40	Indole	Fragrant, jasmine, stables

* Average of triplicate analyses.
SOURCE: Adapted from ref. 22.

In the same study, these flavor researchers quantified the volatile flavor components in pasteurized milk by the Charm technique (22). The results in Table III indicate that of the six key odorants detected, only four were common to both types of milk, but with different proportions and intensities. This demonstrates that significant and measurable flavor differences occur during raw milk pasteurization.

It is interesting to note that a different set of key aromas was established by Ott et al (23) in a mixture of pasteurized full fat milk and skim milk powder. The researchers applied a newly developed GC aromagram technique, SNIF (Surface of Nasal Impact Frequency) to obtain the values for key milk flavor components shown in Table IV.

Table III. Potent Odorants in Pasteurized Bovine Milk

Mean* Charm Value	Compound	Odor Descriptor
2541	Dimethylsulfone	Sulfurous, hot milk, burnt
1601	Hexanal	Freshly cut grass
733	Unknown	Butane, burnt rubber
399	Nonanal	Green, grass-like, tallowy
279	1-Octen-3-ol	Mushroom-like, musty, earth
72	Indole	Fragrant, jasmine, stables

* Average of triplicate analyses.
SOURCE: Adapted from ref. 22.

Table IV. Potent Odorants in Pasteurized Bovine Full Fat / Skim Milk Blend

SNIF Value	Compound	Odor Descriptor
5282	1-Octen-3-one	Mushroom, earthy
5269	Dimethyl sulfide	Milk, lactone-like, sulfury
4470	Hexanoic acid	Rancid, flowery
3487	2-Methylthiophene	Gasoline, plastic, styrene
3421	Pyrazine (?)	
3223	Acetic acid	Pungent, acidic, vinegar
3199	Phenylacetaldehyde	Flowery
3114	1-Nonene-3-one	Mushroom, earthy
3109	Dimethyl trisulfide	Sulfury, H_2S, fecal
3069	2,3-Butanedione	Butter, diacetyl, vanilla
2866	3-Methyl-2-butenal	Metallic, aldehydic, herbaceous

SOURCE: Adapted from ref. 23.

Shiratsuchi et al (24) recently reported C_9-C_{12} alkanoic acids, δ-decalactone, γ- and δ-undecalactone, and γ-dodecalactone as key character-impact compounds responsible for the sweet and milky attributes of skim milk flavor, however no quantitative flavor data were obtained.

Cream Flavor

Cream is produced during the process of removing butterfat from whole milk. The flavor of cream and butter is primarily influenced by compounds present in the lipid fraction of milk, or are formed from precursors during thermal processing (2). Cream flavor differs from that of sweet cream butter because it retains a greater proportion of intact milkfat globule membranes, which comprise an aqueous emulsion of triacylglycerides and phospholipids.

It is interesting to note that a critical organoleptic assessment of cream has not been noted in the literature. While aroma extract dilution studies have not been reported, previous reviews indicate that γ- and δ-decalactones, δ-dodecalactone, and (Z)-4-heptenal, and possibly mercaptans such as hydrogen sulfide and dimethyl sulfide (2,9) are important contributors to cream flavor.

Critical Flavors in Butter

Butter is a water-in-oil emulsion which is formed via the separation of butterfat from the aqueous milk protein phase through churning. The majority of butter manufactured in the United States is obtained from pasteurized fresh sweet cream. The resulting sweet cream butter possesses the unique and desirable characteristics typical of milkfat, but has a bland flavor and does not provide a cultured flavor. Cultured cream butter was routinely made prior to the development of the modern butter manufacturing process. The flavor profile of cultured cream butter contains more of the true cultured notes, which include inherent flavor compounds generated by microbial activity, rather than the use of simulated culturing techniques. More than 230 volatile compounds have been identified in different types of butter and butter oil (1).

The typical flavor of fresh butter is influenced by carbonyl compounds formed via oxidation of unsaturated fatty acids in milkfat. Under severe oxidative conditions or exposure to intense light, it is reported that green, metallic or fishy off-flavors will develop (25). Recently, five different types of butter including sweet cream butter, cultured butter, sour cream butter, and German farm sour cream butter were examined by Schieberle et al (17) for character impact odor compounds. From these studies, it was established that the sensorially important flavor compounds in butter are diacetyl, δ-decalactone, and butyric acid. More recent studies by Budin (26) verify that C_{10}-C_{12} lactones are important contributors to fresh butter flavor, however in their investigation diacetyl and butyric acid were not detected as potent odorants (Table V).

Aroma extract dilution analysis has also been applied to butter oil (27), heated butter (26), and puff pastries prepared with butter (28). Key aroma compounds in heated butter are outlined in Table VI (26). Of interest is that a significant contribution to the flavor of heated butter is due to increases in lactones, unsaturated aldehydes and ketones. Unsaturated triacylglyerides in butterfat are presumed to liberate these potent odorants during heating. Additionally, thermal degradation of proteins can account for formation of skatole and methional, while carbohydrates are precursors of Furaneol™ (4-hydroxy-2,5-dimethyl-2H-furan-3-one).

Table V. Potent Odorants in Fresh Sweet Cream Butter

FD Factor	Compound	Odor Descriptor
512	δ-Decalactone	Coconut-like
256	1-Hexen-3-one	Vegetable-like
256	δ-Dodecalactone	Coconut-like
128	1-Octen-3-one	Mushroom-like
128	Skatole (3-Me indole)	Fecal-like
128	(Z)-6-Dodeceno-γ-lactone	Peach-like

SOURCE: Adapted from ref. 26.

Table VI. Potent Odorants in Heated Butter

FD Factor	Compound	Odor Descriptor
1024	δ-Decalactone	Coconut-like
512	Skatole (3-Me-indole)	Fecal-like
256	Methional	Potato-like
256	δ-Dodecalactone	Coconut-like
256	Furaneol	Burnt sugar
256	1-Octen-3-one	Mushroom-like
128	1-Hexen-3-one	Vegetable-like
128	cis-2-Nonenal	Green, fatty
128	trans-2-Nonenal	Fatty, tallowy
128	trans, trans-2,4-Decadienal	Fatty
128	trans-4,5-Epoxy-trans-2-decenal	Metallic
128	γ-Octalactone	Peach/coconut-like

SOURCE: Adapted from ref. 26.

Critical Flavors in Cultured Dairy Products

The sensory impressions of cultured dairy products such as buttermilk, sour cream and yogurt originate from microbial and enzymatic transformations produced by various culture organisms. These culture-generated flavors are superimposed onto contributions from the native volatile constituents in milk, produced by flavor pathways as previously discussed.

All cultured dairy products receive some degree of heat treatment during their production processes. The severity of heat treatment influences the type and intensity of flavor components that will be generated via Maillard and other related pathways.

Thermally induced compounds contribute to the flavor of cultured products which receive substantial heat treatment.

Buttermilk

Cultured buttermilk is manufactured by fermenting skim milk, whole milk, or reconstituted nonfat dry milk with lactic acid bacteria. Flavor development occurs through the action of mixed strains of lactic *streptococci* (which produce lactic acid) and also *leuconostoc* microorganisms. Aroma-producing starter cultures are responsible for fermentation of citrate into diacetyl, which is paramount in providing the sweet-buttery aroma of fresh buttermilk (29).

Either sour or sweet cream buttermilk can also be obtained as a byproduct of the butter making process. Sweet cream buttermilk is further processed to fermented buttermilk by culturing with lactic acid bacteria. Sour cream buttermilk has a lower flavor stability, and metallic off-flavors can be detected after four days of storage.

In studies of fresh fermented buttermilk flavor by Schieberle *et al.* (25), aroma extract dilution analysis provided 13 compounds with flavor dilution (FD) factors in the range of 8 to 64; seven of these are delineated in Table VII. Of interest is that diacetyl (2,3-butanedione) was perceived among the least potent odorants in buttermilk, even though its predominant aroma is described as "sweet-buttery."

Table VII . Potent Odorants in Fresh Sweet Cream Buttermilk

FD Factor	Compound	Odor Descriptor
64	γ-Decalactone	Coconut-like, sweet
64	δ-Decalactone	Coconut-like, sweet
32	δ-Octalactone	Coconut-like, sweet
32	γ-Octalactone	Coconut-like, sweet
16	Unknown	Goat-like
16	Vanillin	Vanilla
8	2,3-Butanedione	Buttery

SOURCE: Adapted from ref. 25.

A different set of odorants was identified in sour cream buttermilk which developed an intense metallic, cucumber-like sensory defect after four days of storage at 8°C. Of volatiles with FD factors identified in the range 16 to 512, key odor active compounds are presented in Table VIII. The highest FD factor was attributed to (E,Z)-2,6-nonadienol, while other compounds already detected in fresh buttermilk had increased in the stored buttermilk sample.

Table VIII. Potent Odorants in Stored Sour Cream Buttermilk

FD Factor	Compound	Odor Descriptor
512	δ-Decalactone	Coconut-like, sweet
256	(E,Z)-2,6-Nonadienol	Metallic
256	4,5-Epoxy-(E)-2-decenal	Metallic
128	Unknown	Goat-like
128	3-Methyl indole	Mothball
128	Vanillin	Vanilla
64	γ-Octalactone	Coconut-like, sweet
64	γ-Nonalactone	Sweet
64	(E)-2-Undecenal	Green, fatty
64	Methional	Cooked potato

SOURCE: Adapted from ref. 25.

Yogurt

Yogurt exhibits a delicate and low intense flavor which requires mild sample isolation techniques and sensitive identification. Yogurt is typically prepared by inoculation of milk with a starter culture of *Streptococcus thermophilus* and *Lactobacillus bulgaricus*, and its flavor is a combination of compounds already present in milk plus secondary metabolites generated by homofermentative lactic acid bacteria. Much of yogurt flavor is influenced by acetaldehyde; however, recently Imhof (*30*) suggested that 2,3-butanedione, 2,3-pentanedione, dimethyl sulfide, and benzaldehyde were the most potent flavor components. Later Ott *et al* (*23*) reported that, of the 91 volatiles identified by GC/MS in yogurt, 21 exhibited a key aroma impact as determined by SNIF (Surface of Nasal Impact Frequency) values (Table IX), which are relative sensory intensities computed from GC aromagrams. Most recently Ott *et al* (*31*) quantified key impact flavor volatiles in traditional and mild yogurts by GC headspace, and confirmed that yogurt aroma is the superposition of milk odorants on those produced by lactic fermentation.

Critical Flavors in Cheese

Cheese is thought to have originated in southwestern Asia some 8,000 years ago. The Romans encouraged technological improvements and stimulated the development of new varieties during their invasions of Europe between about 60 B.C. and 300 A.D.

Cheese is obtained from curdled milk by separating off the whey. Although there are over 2000 varieties, these represent only 20 basic categories. All cheese

Table IX. Potent Odorants in Yogurt

SNIF Value	Compound	Odor Descriptor
8561	2,3-Butanedione	Butter, diacetyl, vanilla
7006	Acetaldehyde	Fresh, green, pungent
4048	Hexanoic acid	Rancid, flowery
3969	Dimethyl sulfide	Milk, lactone-like, sulfury
3549	2,3-Pentanedione	Butter, vanilla, mild
3486	Benzothiazole	Burnt, rubbery
3377	Guaiacol	Bacon, phenolic, smoked, spicy
2581	Methional	Soup, cooked vegetable, sulfury
2507	Unknown	Metallic
2480	2-Methylthiophene	Gasoline, plastic, styrene
2410	2-Me-tetrahydrothiophene	Green, leather, sulfury
	(E)-2-Nonenal	
2357	1-Octen-3-one	Mushroom, earthy
2339	3-Me-butyric acid	Sweaty, cheese, soy sauce

SOURCE: Adapted from ref. 23.

begins with mammalian milk, usually from a cow. Acid or rennet enzymes coagulate the primary milk protein (casein) into a curd and the free whey (serum portion) is removed. What happens subsequently determines the type and flavor character of cheese produced.

Cheese can be broadly and simply classified into two groups: unripened (fresh) and ripened. Unripened cheese is made from milk coagulated by acid or high heat and must be consumed shortly after it is made or it will spoil. Cottage cheese is the most familiar example, but the category also includes cream cheese, Neufchatel, ricotta and mozzarella. The unripened cheeses exhibit a mild delicate flavor that is derived from three sources; the milk, heat degradation products of milk, and the metabolic products of microorganisms used as cultures (*32, 33*).

Ripened cheese is made from milk fermented by lactic acid bacteria and coagulated by a rennet-like enzyme preparation. The curd is pressed to remove the whey and is salted; then it is held for an extended period in a controlled environment. During that time various physical and chemical changes take place to give the material the characteristic flavor and texture that cause it to be classified as a specific variety of cheese. The great majority of cheeses are ripened by the addition of selected microorganisms. Ripening cheese represents a dynamic system in which many reactions are proceeding simultaneously. Many of the reactions are mediated through fermentation pathways of the microorganisms, through enzymes from lysed microbial cells, and through enzymes native to milk. In the course of these reactions the protein, fat, and carbohydrate are degraded to varying degrees to yield a complex mixture of flavoring compounds.

Classification of Cheeses (*34*)

    I) Natural
        A) Unripened
            1) *Low Fat-* Cottage, Baker's
            2) *High Fat-* Cream, Neufchatel
        B) Ripened
            1) *Hard Grating Cheese-* Romano, Parmesan, Asiago Old
            2) *Hard-* Cheddar, Swiss, Gruyere, Provolone, Gouda
            3) *Semi-Soft-* Roquefort, Blue, Gorgonzola, Brick, Limburger
            4) *Soft-* Camembert, Brie, Liederkranz
    II) Whey Cheeses
            1) Ricotta
            2) Mysost

Source of Milk

Although cows provide most of the milk for cheesemaking in the United States, many parts of the world rely more on other mammals. In southwestern Asia and along the Mediterranean, sheep and goats constitute the major source. France has more than one million milking goats and large numbers of sheep whose milk goes mostly into the manufacture of Roquefort cheese. Among the animals providing milk for cheese in other parts of the world are water buffalo, camels, yaks, reindeer and llamas. The mammalian origin of the milk influences the flavor and aroma of a natural ripened cheese. Goats' milk gives cheese a spicier and more piquant flavor than cows' milk primarily because its fat has a greater preponderance of C_6, C_8, and C_{10} fatty acids. Each acid contributes a different pungency to the milk. Sheep's milk yields cheese of a distinctive flavor because the milk contains very significant levels of higher fatty acids. Table X presents the short-chain fatty acid composition of milk commonly used in cheeses.

Table X. Relative Proportion of Fatty Acids

Species	Fatty Acid Ratio				
	C_4	C_6	C_8	C_{10}	C_{12}
Cow	1	1	1	1	1
Buffalo	1	1	0.84	0.63	0.64
Sheep	1.20	1.75	2	3	1.75
Goat	0.78	1.80	2	2.80	1.06

SOURCE: Adapted from ref. 35.

Little of this is noticeable in the taste of the fresh milk. The characteristic flavors appear only when a young cheese made from one of these milks is ripened and fatty acids are released from the triglycerides by lipase enzymes.

Role of Microorganisms

The ripening of cheese is due to a high concentration of microorganisms. On the first day of production the microbial count in the starting material ranges from one to two billion per gram. Thereafter the population declines because of insufficient oxygen, high acidity, depletion of readily fermented carbohydrates and the presence of inhibitory compounds that are produced as the cheese ripens.

Fortunately, the ripening organisms are non-pathogenic and safe for human consumption. There is mounting evidence that many are beneficial to overall health (36). It is largely the action of cellular enzymes on lactose, fat and protein in milk that creates ripened-cheese flavor.

Cheese Flavor

The estimated thousands of cheese varieties available throughout the world provide a broad spectrum of flavors. All varieties are made from the milk of mammals, so species determines some varietal differences. In addition, the manner in which the milk is manipulated to form curds and the subsequent dehydration and compaction of the curds into the final shape of the cheese, the normal and added microflora, and the procedure employed to ripen the cheese also are factors in determining varietal differences.

The studies discussed below summarize results gathered from recent studies where the authors have emphasized sensory experiments to correlate measurable responses to GC separated components to determine the key aroma compounds. Such techniques include CharmAnalysis, AEDA, OAV, FD studies etc. as discussed in a previous section.

Specific Cheese Studies

Camembert

Camembert is an externally mold ripened French cheese which esessentially the same as Brie. It has a soft and buttery texture with a salty, slightly rancid, buttertaste. Whole milk, (sometimes with cream added) is inoculated with *Lactococcus* sp. to develop lactic acid, followed by a coagulation step. The surface ripening is due to the mold *Penicillium camemberti*. Total ripening time is 4 to 6 weeks. The proteolytic and lipolytic enzymes of the mold are responsible for many of the flavor compounds of this cheese (*37*).

Recently, the potent odorants of Camembert cheese were screened and evaluated by AEDA, and quantified by stable isotope dilution assays (*38,39*). Camembert cheese was frozen in liquid nitrogen and ground. The powder was mixed with sodium sulfate and extracted with diethyl ether. The ether was filtered and concentrated. Finally, the organic residue was distilled under high vacuum. The condensate was separated into neutral and acid fractions by sodium carbonate extraction. The basic aqueous fraction was made acid and re-extracted with diethyl ether to produce the organic acid fraction.

To assess the contribution of the most volatile components, a gas chromatographic-olfactometric assay was performed on a static headspace sample. In this case, a sample of ground Camembert was held in a water bath at 25°C, and 40-mL samples of headspace were injected.

AEDA analysis was performed on the neutral volatile diethyl ether fraction and 46 peaks were observed. A summary of the most potent compounds is given in Table XI. A flavor model based on an unripened, freeze-dried cheese of the Mozzarella type approached the sensory properties of Camembert by the addition of these compounds, which had been screened as important odor and taste contributors (*40*).

Table XI. Potent Odorants of Camembert Cheese

FD Factor	Compound	Odor Descriptor
512	3-Methylbutanal	Green, malty
512	Methional	Boiled potato-like
256	2-Undecanone	Floral
128	2,3-Butanedione	Buttery
128	1-Octen-3-one	Mushroom-like
128	1-Octen-3-ol	Mushroom-like
128	δ-Decalactone	Coconut-like

SOURCE: Adapted from ref. 38.

In addition, the authors also performed AEDA on the acid fraction and reported that butyric and isovaleric were the important acidic odorants of the cheese. Other acids were of only minor importance.

Finally, they performed a gas chromatography-olfactometry assay of a static headspace sample. In this fashion, they were able to demonstrate the importance of methanethiol and dimethyl sulfide as additional important high volatile components of Camembert cheese.

In their discussion, they stressed the importance of 2,3-butanedione, 3-methylbutanal, methional, 1-octen-3-one, 1-octen-3-ol, 2-undecanone, δ-decalactone, butyric and isovaleric acids, as well as the sulfur compounds methanethiol and dimethyl sulfide as important contributors of the aroma. They noted that the two 1-octen-3-ol and -one compounds are responsible for the characteristic mushroom note of Camembert.

Goat Cheese

Cheeses made from goats' milk are generally described as possessing a strong, typical flavor. This goaty character originates from the lipid fraction. Generally branched chain fatty acids of 8-10 carbon atoms are considered to be aroma contributors of goat cheeses; fatty acids branched at the 4-position are described as "goaty, muttony, sheepy" in character.

Very recently, French researchers studied the character impact compounds of a traditional local cheese ('Bouton de Culotte' de Salon et Loire) by gas chromatography-olfactometry (41). 'Bouton de Culotte' cheese was frozen and homogenized with water and centrifuged. The process was repeated on the pellet. The aqueous supernatants were combined, centrifuged again and extracted with methylene chloride. Odor assessment of this material demonstrated a strong, characteristic goat cheese character.

This sample was submitted to CharmAnalysis using a panel of 3 trained assessors. Six peaks with Charm values of 1024 or greater were identified. The results of that assay are presented in Table XII.

Of significance is the fact that two of the more potent compounds in this series are branched fatty acids. Other compounds with lower CharmAnalysis values were butanoic, hexanoic, octanoic, nonanoic and decanoic acids. The authors observed that this study confirmed the importance of branched fatty acids to goat cheese aroma.

Table XII. Potent Odorants of Goat Cheese

Charm Value	Compound	Odor Descriptor
2048	Methional	Potato
2048	4-Ethyl octanoic acid	Goat
1024	3-Methyl butanoic acid	Cheese
1024	Phenylethanol	Rose
1024	δ-Octalactone	Fruity
1024	Nonanoic acid	Goat

SOURCE: Adapted from ref. 41.

As a result of these studies the authors prepared a synthetic mixture of the major fatty acids and added these to a flavorless, tasteless model cheese. The overall "goat" character was completely explained by the branched chain fatty acids.

Cheddar Cheese

The flavor of cheddar cheese is described as sweet, buttery aromatic, and walnut with no outstanding single note. In aged cheese, a biting quality gives sharpness to the cheese. Cheddar is consumed when it is from one month to several years old. Whole milk is and *Lactococcus lactis* subsp *cremoris* is typically used as the starter. Only cheese with 50% fat develops a typical flavor.

Several pertinent studies have been published since 1995. Cheddar cheese headspace was evaluated by GC sniffing techniques by Arora et al (*42*). Cheddar cheese and sodium sulfate were ground together and the volatiles were swept and concentrated on a Tenax trap. The trap was thermally desorbed onto a multidimensional GC equiped with both polar and non-polar megabore columns. Panelists were asked to assess the character and intensity of the odor-active components.

The four compounds with greatest intensity reported in this study were 2-butanone, 2,3-butanedione, ethyl butanoate and 3-hydroxy-2-butanone. The authors concluded that lipid derived aldehydes, methyl ketones and esters were the principal aroma-bearing components. They observed that further studies were required to relate concentration of the components to threshold level and to observe any synergistic effects.

A 1995 publication by Christensen and Reineccius (*43*) analyzed aged Cheddar cheese by AEDA. Commercial aged Cheddar cheese was cubed and centrifuged at 40°C. The recovered oil was subjected to molecular distillation at 50°C. The condensate was extracted with diethyl ether, dried and concentrated. Acids were not separated from the neutral components.

The AEDA experiment was performed on successive samples which were diluted sequentially in a 1:3 ratio. Evaluation was done by two experienced subjects working independently. The results of the experiment are presented in Table XIII.

From this data, the importance of butanoic acid as well as several other fatty acids to sharp Cheddar flavor is evident. Also significant to the flavor are ethyl esters, as well as methional with its cheesy, potato character. The authors concluded that AEDA analysis is helpful to narrow the group of compounds that are important to Cheddar cheese aroma. They observe that their procedure does not permit detection of highly volatile compounds such as methyl mercaptan or hydrogen sulfide, and further note that such compounds may play an important role in Cheddar flavor.

The same University of Minnesota group quantified the potent odorants of mild cheddar cheese in 1997 (*44*). The frozen cheese was extracted with diethyl ether, the ether dried and distilled-off, and finally the residue was subjected to high vaccuum

Table XIII. Potent Odorants of Sharp Cheddar Cheese

FD Value	Compound	Odor Descriptor
729	Butanoic acid	Nacho cheese
243	Ethyl butyrate	Sweet fruity
81	Ethyl caproate	Fruit
81	2- and 3-Methyl butanal	Tootsie Rolls
81	Propionic acid	
81	Hexanoic acid	Goaty
27	Acetic acid	
27	Pentanoic acid	
9	1-Octen-3-one	Mushroom
9	Methional	Cooked milk
9	Octanoic acid	Stale butter
9	Decanoic acid	Soapy

SOURCE: Adapted from ref. 43.

distillation. The product had a butter-like, sweet-fruity acidic aroma. An AEDA experiment was performed on both the distillate and a static headspace sample, and the results are presented in Table XIV. They concluded the above compounds possess a high aroma impact, and also observed that, among the highly volatile compounds, methanethiol and dimethyl sulfide are important to nasal perception.

Table XIV. Potent Odorants of Mild Cheddar Cheese

FD Value	Compound	Odor Descriptor
512	5-Ethyl-4-hydroxy-2-methyl-2 H-furan-3-one	Caramel
512	Butanoic acid	Sweet, sweaty
128	4-Hydroxy-2,5-dimethyl-2 H-furan-3-one	Caramel
128	δ-Decalactone	Coconut-like
64	Skatole	Fecal
64	6-(Z)-Dodecen-γ-lactone	Soapy
32	Methional	Boiled potato
32	(E)-β-Damascenone	Fruity, peach-like
16	Acetic acid	Pungent
16	Diacetyl	Butter-like

SOURCE: Adapted from ref. 44.

As a continuation of the same study, OAV's were calculated (retronasal, in oil) to yield the results in Table XV. They concluded that these volatile compounds are the primary odor contributors to the mild, pleasant aroma of Cheddar cheese. It should be noted that nasal OAV's in water and oil were also reported.

Table XV. OAV of Potent Odorants in Mild Cheddar Cheese

OAV	Compound
60	Methional
36	5-Ethyl-4-hydroxy-2-methyl-2 *H*-furan-3-one
21	Diacetyl
15	Butanoic acid
14	Acetic acid
12	4-Hydroxy-2,5-dimethyl-2 *H*-furan-3-one

SOURCE: Adapted from ref. 44.

Mozzarella

Water buffalo mozzarella is a soft plastic-curd cheese that is made in southern Italy. Originally it was made only from buffalos' milk, but now is made also from cows' milk. The culture used in its preparation is a combination of *Streptococcus thermophilus* and either *Lactobacillus delbrueckii* or *Lb. helveticus* (*45*). The cheese is typically aged for a few days.

Although no systematic study has been made on the characterizing components of Mozzarella, one recent study is of interest. A series of 1993/1994 publications studied Mozzarella cheese and the authors detected a GC-eluting compound with a melted cheese flavor (*46,47*).

Mozzarella cheese from both bovine and water buffalo milk was homogenized and vacuum distilled. The aqueous phase was extracted with methylene chloride and concentrated. The sample was separated by gas chromatography (*46*). One component was detected with an odor of melted cheese. In a subsequent publication, the authors studied the formation of an odor-characterizing compound in acidified water buffalo milk (*47*). They vacuum-distilled the milk and submitted the condensate to methylene chloride extraction. One major compound was detected, and spectroscopic analysis identified it as 3-hydroxy-5-methyl-2-hexanone, which posessed an odor of melted cheese.

Swiss Cheese

Swiss-type cheeses (Emmental, Swiss, Gruyere) are hard cheeses characterized by an elastic body, smooth texture, uniform smooth eyes and a unique nut-like flavor. Eyes develop through formation of CO_2 by bacteria such as *Propionibacterium* sp. It is the development of propionic acid during fermentation that distinguishes Swiss from all other cheeses. Eye development is key, and may be the only criteria by which these cheeses are graded. Starters for Swiss are predominately *Streptococcus thermophilus* and either *Lactobacillus delbrueckii* or *Lb. helveticus* (*48*).

A series of publications have studied the potent odorants and character impact compounds of Swiss cheese. One study investigated the volatile neutral compounds of Swiss cheese. Emmentaler cheese was frozen, ground and extracted with diethyl ether. The volatiles were separated from the residue by distillation, and the acids removed by bicarbonate extraction. The sample was separated by gas chromatography and potent odorants located by aroma extract dilution analysis (AEDA). The results of that study are presented in Table XVI (*49*); the authors suggest that these compounds are responsible for the sweet, mild nut-like aroma of Swiss cheese.

Table XVI. Potent Odorants of Swiss (Emmentaler) Cheese

FD Value	Compound
512	δ-Decalactone
512	(Z)-6-Dodecen-γ-lactone
256	Skatole
128	5-Ethyl-4-hydroxy-2-methyl-2*H*-furan-3-one
128	4-Hydroxy-2,5-dimethyl-2*H*-furan-3-one
128	2-*sec*-Butyl-3-methoxy pyrazine
128	Ethyl butanoate

SOURCE: Adapted from ref. 49.

In a subsequent publication these authors calculated Odor Activity Values (OAV's) to determine the key odorants (*50*). Again, Emmentaler cheese was frozen, ground and extracted with diethyl ether. The volatiles were separated from the residue by distillation, and the acids removed by bicarbonate extraction. The results of the previous AEDA study were used and thirteen potent odorants were quantified by isotope dilution asssay. The odor activity values (OAV's) were calculated by dividing the concentrations of the odorants in the cheese by the nasal odor threshold in oil. Oil was selected as the evaluation media since the fat phase was the predominating solvent in these cheeses. The results are presented in Table XVII.

Table XVII. Key Odorants of Swiss (Emmentaler) Cheese

Odor Activity Value[1]	Compound
121	Methional
57	4-Hydroxy-2,5-dimethyl-2*H*-furan-3-one
30	5-Ethyl-4-hydroxy-2-methyl-2*H*-furan-3-one
14	3-Methyl butanal
5	Ethyl 3-methyl butanoate
5	2,3-Butanedione
3	Ethyl hexanoate

[1]Average of two samples
SOURCE: Adapted from ref. 50.

They concluded that methional, 4-hydroxy-2,5-dimethyl-2*H*-furan-3-one (Furaneol™) and 5-ethyl-4-hydroxy-2-methyl-2*H*-furan-3-one (homofuraneol) are the key compounds to the flavor character of Emmentaler cheese and that the furanones are responsible for the sweet note. The difference between the FD-factor results and the OAV values was attributed to the fact that great amounts of methional and the furanones were lost during workup of the sample for the AEDA study. Losses were compensated for during the OAV study by the addition of internal standards labeled with stable isotopes (*51*).

To confirm the importance of the above, Preininger *et al* (*52*) prepared flavor systems using the above results. The flavor models were added to a bland base which was an unripened, freeze dried Mozzarella. Models composed of methional, 4-hydroxy-2,5-dimethyl-2*H*-furan-3-one, 5-ethyl-4-hydroxy-2-methyl-2*H*-furan-3-one, acetic, propionic, lactic, succinic and glutamic acids and inorganic salts were judged to closely match the flavor of Swiss cheese.

Internally Ripened Blue Mold Cheeses

Although no contemporary odor assessment studies have been done on this group of cheeses, their characteristic flavor and the fact that they have been studied so comprehensively suggest that they be included in this survey.

Roquefort, Stilton, blue and Gorgonzola are cheeses in which development of the peppery, piquant flavors is due to metabolism of *Penicillium roqueforti* or *Penicilliuin glaucum*. These molds grow throughout the cheese and are able to grow in the low-oxygen, high-salt conditions that are typical of these cheeses.

Roquefort cheese must be made from whole sheeps' milk and ripened in high-humidity caves near the town of Roquefort, France. Similar cheese produced from cows' milk is called blue (bleu) cheese. Blue cheese is usually made from a blend of skim milk and homogenized cream. The starter is *Lactococcus lactis,* and citrate-

Table XVIII. Important Flavor Compounds in Selected Cheeses

Cheese	Compounds
Cheddar	Methanethiol
Camembert	Nonan-2-one, oct-1-ene-3-ol, N-isobutylacetamide, 2-phenylethanol, 2-phenylethyl acetate, 2-heptanol, 2-nonanol, NH_3, isovaleric acid, isobutyric acid, hydroxybenzoic acid, hydroxyphenylacetic acid
Emmental	Methional, 4-hydroxy-2,5-dimethyl-3(2H)-furanone (Furaneol) and 5-ethyl-4-hydroxy-2-methyl-3(2H)-furanone
Romano	Butanoic, hexanoic, and octanoic acid
Parmesan	Butanoic acid, hexanoic acid, octanoic acid, ethyl butyrate, ethyl hexanoate, ethyl acetate, ethyl octanoate, ethyl decanoate, methyl hexanoate
Provolone	Butanoic, hexanoic, and octanoic acid
Goats' milk cheese	4-Methyloctanoic acid, 4-ethyloctanoic acid
Sheeps' milk romano cheese	4-Methyloctanoic and 4-ethyloctanoic acids, p-cresol, m-cresol, 3,4-dimethylphenol
Limburg	Methanethiol, methyl thioacetate
Surface ripened cheeses	Methyl thioesters
Pont-l'Eveque	N-Isobutylacetamide, phenol, isobutyric acid, 3-methylvaleric acid, isovaleric acid, 2-heptanone, 2-nonanone, acetophenone, 2-phenylethanol, indole
Vacherin	Acetophenone, phenol, dimethyl disulfide, indole, terpineol, isoborneol, linalool
Roquefort	Oct-1-en-3-ol, methyl ketones
Livarot	Phenol, m- and p-cresol, dimethyl disulfide, isobutyric acid, 3-methylvaleric acid, isovaleric acid, benzoic acid, phenylacetic acid, 2-nonanone, acetophenone, 2-phenylethanol, 2-phenylethyl acetate, dimethyl disulfide, indole
Munster	Dimethyl disulfide, isobutyric acid, 3-methylvaleric acid, isovaleric acid, benzoic acid, phenylacetic acid

Trappiste	H_2S, methanethiol
Blue cheeses	2-Heptanone, 2-nonanone, methyl esters of $C_{4,6,8,10,12}$ acids, ethyl esters of $C_{1,2,4,6,8,10}$ acids
Brie	Isobutyric acid, isovaleric acid, methyl ketones, sulfur compounds, oct-1-en-3-ol
Epoisses	2-Phenylethanol
Buffalo mozzarella	Oct-1-en-3-ol, nonanal, indole
Cow mozzarella	Ethyl isobutanoate, ethyl 3-methylbutanoate

SOURCE: Adapted from ref. 9.

metabolizing strains are sometimes added. The carbon dioxide produced through metabolism of citric acid expands openings in the cheese, which, in turn, allows for more intrusive growth of the mold.

Cheeses are brine salted or rubbed with salt then pierced with needles and placed in a curing room for 2 to 4 weeks until mold growth begins to appear at the openings of the holes. The piercing allows for transfer of oxygen and CO_2 to stimulate growth and metabolism of *P. roqueforti*. After sufficient mold growth, cheeses are wrapped and stored (matured) for 2 to 4 months.

Metabolism of the molds (lipolysis and proteolysis) during maturation is essential for development of the distinctive blue cheese flavor. Mold ripened cheeses have a high free fatty acid content. The acids are released through lipolysis and are converted to methyl ketones via oxidative decarboxylation. Of particular importance in blue-veined cheeses are 2-heptanone and 2-nonanone which are critical to blue cheese flavor. Secondary alcohols, methyl and ethyl esters derived from fatty acid metabolism and proteolysis, are essential for well-balanced and distinctive flavor of blue veined cheeses (*53*).

Qualitative Assessment of Important Cheese Compounds

The previous studies were those where recent research reported analytical techniques to measure the odor active volatiles. In a 1997 publication, Urbach (*9*) reported on the flavor of milk and dairy products. The author discussed the formation of volatile compounds and listed important flavor compounds in different varieties of cheeses. A portion of that compilation is presented in Table XVIII and represents a contemporary, qualitative assessment of the important aroma compounds of cheeses.

Conclusion

While the aromatic composition of dairy products has been studied since the mid- 1950's, the great majority of these works have presented the qualitative and quantitative composition, but not the relevance of the identifications.

The objective of this chapter was to critique the recent published technical literature with a goal of providing a review of the critical flavor compounds of cow's milk, sour cream, yogurt and various cheeses. Particular emphasis was placed on studies involving organoleptic-gas chromatographic assays such as AEDA, OAV and flavor dilution factor studies. The data we have summarized can be used in creative flavor compounding efforts to replicate natural flavors as well as to monitor production of cheese and dairy flavors.

Although the work presented here represents the current state of the art of critical compound flavor analysis as of 1998, one further aspect remains absent. Only a few of the researchers (*40,41,52*) have applied their data and used it to replicate the analyzed dairy flavor in a model system. Such studies are needed to confirm the actual criticality of the identified compounds.

Acknowledgments

The authors wish to thank Sara McGarty, Technical Information Services, Kraft Foods, for assistance with the literature search and Martin Preininger, Kraft Foods, for valuable discussions and suggestions.

References

1. Maarse, H.; Visscher, C. A.; Willemsens, L. C.; Nijssen, L. M.; Boelens, M. H., Eds., *Volatile Compounds in Food, Qualitative and Quantitative Data*, Supplement 5 to the 6th ed., TNO Nutrition and Foods Research: Zeist, Netherlands, 1994.
2. Badings, H. T.; Neeter, R, *Neth. Milk Dairy J.* **1980**, *34*, 9-30.
3. Badings, H. T. In *Volatile Compounds in Foods and Beverages*, Maarse, H., Ed., Marcel Dekker: New York, NY, 1991, pp. 91-106.
4. Imhof, R.; Bosset, J. O., *Zeitschrift fur Lebensmittel-Untersuchung und Forschung* **1994**, *198*, 267-276.
5. Adda, J. In *Developments in Food Flavors*, Birch, G. G.; Lindley, M. G., Eds., Elsevier Applied Science: London, 1986, pp. 151-172.
6. Hammond, E. G. In *Flavor Chemistry of Lipid Foods*, Min, D. B.; Smouse, T. H., Eds., American Oil Chemists Society: Champaign, IL, 1989, pp. 222-236.
7. McSweeney, P. L. H.; Nursten, H. E.; Urbach, G. In *Advanced Dairy Chemistry–Lactose, Water, Salts and Vitamins*, Fox, P. F., Ed., Chapman and Hall: London, 1997, pp. 403-468.
8. Nursten, H. E., *Int. J. Dairy Technol.* **1997**, *50*, 48-56.
9. Urbach, G. *Int. J. Dairy Technol.* **1997**, *50*, 79-89.
10. Marsili, R., Ed. *Techniques for Analyzing Food Aroma*, Marcel Dekker, Inc.: New York, NY, 1997.
11. Teranishi, R.; Kint, S. In *Flavor Science: Sensory Principles and Techniques*, Acree, T. E.; Teranishi, R., Eds., ACS Professional Reference Book, American Chemical Society, Washington, DC, 1993, pp 137-167.
12. Chaintreau, A.; Maignial, L.; Pollien, P.; "Apparatus and process for simultaneous distillation and extraction", Eur. pat. 95 203 465.5, 1995.
13. Schieberle, P. In *Characterization of Food: Emerging Methods,* Gaonkar, A. G., Ed., Elsevier Applied Science: Netherlands, 1995, pp. 403-431.
14. Acree, T. E. In *Flavor Science: Sensible Principles and Techniques*, Acree, T. E.; Teranishi, R., Eds., ACS Professional Reference Book, American Chemical Society: Washington, DC, 1993, pp. 1-20.
15. Schieberle, P.; Grosch, W. *Z. Lebensm. Unters. Forsch.* **1987**, *185*, 111-113.

16. Blank, I.; Sen, A.; Grosch, W. *Z. Lebensm. Unters. Forsch.* **1992**, *195*, 239-245.
17. Schieberle, P.; Gassenmeier, K.; Guth, H.; Sen, A.; Grosch, W. *Lebensm. Wiss. Technol.* **1993**, *26*, 347-356.
18. Stahl, W. H., Ed. *Compilation of Odor and Taste Threshold Values Data*, American Society for Testing and Materials: Philadelphia, PA, 1973.
19. Pollien, P.; Ott, A.; Montigon, F.; Baumgartner, M.; Munoz-Box, R.; Chaintreau, A. *J. Agric. Food Chem.* **1997**, *45*, 2630-2637.
20. Calvo, M. M.; de la Hoz, L. *Int. Dairy Journal* **1992**, *2*, 69-81.
21. Moio, L.; Langlois, D.; Etievant, P.; Addeo, F. *J. Dairy Research* **1993**, *60*, 215-222.
22. Moio, L.; Etievant, P.; Langlois, D.; Dekimpe, J.; Addeo, F. *J. Dairy Research* **1994**, *61*, 385-394.
23. Ott, A.; Fay, L. B.; Chaintreau, A. *J. Agric. Food Chem.* **1997**, *45*, 850-858.
24. Shiratsuchi, H.; Yoshimura, Y.; Shimoda, M.; Noda, K.; Osajima, Y., *J. Agric. Food Chem.* **1995**, *43*, 2453-2457.
25. Heiler, C.; Schieberle, P. *Lebensm. Wiss. Technol.* **1996**, *29*, 460-464.
26. Budin, J., Ph.D. Thesis, Univ. of Minnesota, Minneapolis, MN, 1998.
27. Widder, S.; Sen, A.; Grosch, W. *Z. Lebensm. Unters. Forsch.* **1991**, *193*, 32-35.
28. Gassenmeier, K.; Schieberle, P. *Lebensm. Wiss. Technol.* **1994**, *27*, 282-288.
29. Marshall, V. M. *J. Soc. Dairy Technol.* **1993**, *46*, 49-56.
30. Imhof, R.; Glatti, H.; Bosset, J. O. *Lebensm.-Wiss. Technol.* **1995**, *28*, 78-86.
31. Ott, A.; Germond, J.-E.; Baumgartner, M.; Chaintreau, A. *J. Agric. Food Chem.* **1999**, *47*, 2379-2385.
32. Kosikowski, F. *Scientific American* **1985**, *252* (5), 88-99.
33. Nauth, K. R.; Hynes, J. T.; Harris, R. D. In *Encyclopedia of Food Science and Technology,* John Wiley & Sons Inc.: New York, NY, 1991, pp 333-348.
34. Day, E. In *The Chemistry and Physiology of Flavors*, Schultz, H.; Day, E.; Libbey, L., Eds., AVI Pub Co.: Westport, CT, 1967, pp 331-361.
35. Fox, P. F. *Advanced Dairy Chemistry, Vol 2: Lipids*, 2nd. ed, Chapman & Hall: New York, NY, 1995, p. 13.
36. Salminen, S.; von Wright, A., Eds., *Lactic Acid Bacteria in Health and Disease*, 2nd ed., Marcel Dekker: New York, NY, 1998, pp. 211-254.
37. Margalith, P. Z. *Flavor Microbiology*, Charles C. Thomas Pub.: Springfield, IL, 1981, pp. 90-93.
38. Kubickova, J.; Grosch, W. *Int. Dairy Journal* **1997**, *7*, 65-70.
39. Kubickova, J.; Grosch, W. *Int. Dairy Journal* **1998**, *8*, 17-23.
40. Kubickova, J.; Grosch, W. *Int. Dairy Journal* **1998**, *8*, 11-16.
41. Le Quere, J.L.; Septier, C.; Demaizieres, D.; Salles, C. In *Flavor Science, Recent Developments*, Taylor, A. J.; Mottram, D. S., Eds., The Royal Society of Chemistry: Cambridge, UK, 1997, pp. 325-330.
42. Arora, A. G.; Cormier, F.; Lee, B. *J. Agric. Food Chem.* **1995**, *43*, 748-752.
43. Christensen, K.; Reineccius, G. *J. Food Sci.* **1995**, *60*, 218.
44. Milo, C.; Reineccius, G. *J. Agric. Food Chem.* **1997**, *45*, 3590-3594.

45. Johnson, M. In *Applied Dairy Microbiology*, Marth, E.; Steele, J. Eds., Marcel Dekker: New York, NY, 1998, pp. 234-235.
46. Moio, L.; Dekimpe, J.; Etievant, P.; Addeo, F. *Ital. J. Food Sci.* **1993**, *5*, 215-225.
47. Moio, L.; Semon, E.; LeQuere, J. *Ital. J. Food Sci.* **1994**, *6*, 441-447.
48. Johnson, M. In *Applied Dairy Microbiology*, Marth, E.; Steele, J., Eds., Marcel Dekker: New York, NY, 1998, pp. 229-232.
49. Preininger, M.; Rychlik, M.; Grosch, W. In *Trends in Flavor Research*, Maarse, H.; van der Heij, D., Eds. Elsevier: Amsterdam, Netherlands, 1994, pp. 267-270.
50. Preininger, M.; Grosch, W. *Lebensm. Wiss. u Technol.* **1994**, *27*, 237-244.
51. Grosch, W. *Flavor Frag J.* **1994**, *9*, 147-158.
52. Preininger, M., Warmke, R., Grosch, W. *Z. Lebensm. Unters Forsch.* **1996**, *202*, 30-34.
53. Kinsella, J.; Hwang, D., *Biotechnol. Bioeng.* **1976**, *18*, 927.

Chapter 5

Biosynthesis of Plant Flavors: Analysis and Biotechnological Approach

Wilfried Schwab

Lehrstuhl für Lebensmittelchemie, Universität Würzburg,
Am Hubland, D–97074 Würzburg, Germany

The increasing demand of the consumers for natural food and biological production of food ingredients has intensified the interest of the food industry in the elucidation of the biosynthetic pathways of plant flavors. The knowledge of the biogenesis of plant volatiles, their enzymes and genes will pave the way for the economic biotechnological production of food flavors by plant tissue cultures, micro-organisms and enzymes. Several genes have already been cloned which code for enzymes involved in the biosynthesis of flavor molecules. Among these are a fatty acid hydroperoxide lyase from banana plants (*Musa sp.*) responsible for the formation of „green note" flavors, monoterpene cyclases from *Mentha* species and linalol synthase from *Clarkia breweri*. In contrast to the pharmaceutical industry, the applications of modern biotechnology in the flavor industry are still limited. However, examples such as vanillin, (E)-3-hexenal, lactones, furanones and 1,3-dioxanes show the potential of the new technologies. The current analytical methods for the elucidation of biosynthetic pathways and the opportunities for their biotechnological production are presented.

In recent years the consumers` demand for natural food has increased continuously. This trend can be attributed to increasing health- and nutrition-conscious lifestyles. The consumers usually believe that natural material including flavor are more healthy and safer than the synthetic counterpart. Although scientific evidence does not support this view the consumers belief is very strong. As a consequence the demand for natural ingredients has risen from 10 % of the food company requests to 80 % within the last decade. However, the sources for natural flavor are limited and as the price depends on the supply and the demand costs of natural flavors are immense. The comparison of the costs for natural flavor compounds and their synthetic counterparts show that the chemically produced compounds are by a factor 100-400 cheaper than the natural ones. This price difference justifies the intensive research of several companies and universities for new sources of natural flavors.

In the US natural flavors are defined as: „the essential oil, oleoresin, essence or extractive, protein hydrolysate, distillate, of any product of roasting, heating or enzymolysis, which contains the flavoring constituents derived from a spice, fruit juice, vegetable or vegetable juice, edible yeast, herb, bud, bark, root, leaf or similar plant material, meat, seafood, poultry, eggs, dairy products or fermentation products thereof whose significant function in food is flavoring rather than nutrition" [1]. This means that extracts from natural sources as well compounds produced by biosynthetic processes may be considered natural. All other substances are labeled artificial.

Three flavor categories exist in Europe: Natural, nature-identical and artificial flavors. The definition of naturals is almost identical to the US guidelines: „Materials (mixture or single substances) are called natural if they are obtained exclusively by physical, microbiological, or enzymatic processes from material of vegetable or animal origin, either in the raw state or after processing for human consumption by traditional food preparation processes (including drying, roasting, and fermentation)" [2].

Biotechnologically produced flavors are also covered by the term natural. However, in Europe the term artificial is further subdivided into nature-identical and artificial. A synthetic compound is considered nature-identical if it is identical to the same compound found in nature. The term artificial is reserved for those synthetic components that are not found in nature. Currently, the label „flavoring" is used for nature-identical or artificial compounds. Therefore, the major advantage of biotechnologically produced products is attainment of the natural status and the ability to make such a claim on the product label and the ingredient list.

Plants used for the production of flavors can either be manipulated by conventional plant breeding methods such as intraspecific crossing, hybridization and nonspecific mutagenesis by chemicals or irradiation or by novel plant breeding methods such as tissue culture techniques, protoplast fusion techniques and recombinant DNA techniques [1]. The goals of these expensive breeding strategies are enhanced flavor production, higher extraction yields or disease resistance. An alternative way to this time- and cost-consuming approach would be the biotechnological production of flavors. Basically three sources for the biotechnological production of flavors are available (Figure 1).

These are plant cells, enzymes, and microorganisms [3]. Plant cells and microorganisms have the advantage that they can use relatively inexpensive substrates such as carbohydrates and amino acids to form complex flavor mixtures. However, the concentration of the desired products by *de-novo* synthesis are rather low and the cultures grow slowly. Thus, screening for highly productive strains and genetic engineering is necessary to obtain reasonable amounts. In some cases a simple biotransformation step can form a highly appreciated flavor from a relative cheap starting material. This kind of reaction can be performed by enzymes, microorganisms, and plant cells. The use of enzymes on an industrial scale is common practice now. The enzyme transformations produce extremely pure products, only catalytic amounts are required, and they are extremely selective. However, of the 1500 chemicals that

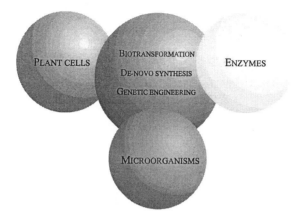

Figure 1: Sources for the biotechnological production of flavors.

are used by the US flavor industry only 20 have been produced commercially by fermentation routes.

A great deal more must be learned about the biochemical and genetic regulations of plant secondary metabolites before large-scale production becomes a commercial reality.

Hexenals

A good example for high-value low-volume products are the leaf aldehyde (E)-2-hexenal) and (Z)-3-hexenal which are responsible for the green flavors and aromas of fruits and vegetables. Currently, synthetic compounds are used extensively. The natural compounds are obtained primarily from plant tissue that have been disrupted in some fashion.

In general, the unsaturated fatty acids linoleic and linolenic acid are degraded via a lipoxygenase-catalyzed formation of hydroperoxides and a subsequent cleavage by a hydroperoxide lyase to form aliphatic C_6-compounds such as (Z)-3-hexenal (Figure 2). The aldehyde is further reduced by alcohol dehydrogenase to (Z)-3-hexenol or isomerized to (E)-2-hexenal and then reduced to the alcohol.

Recently, it was found that the hydroperoxide represents a branching point in the fatty acid metabolism in plants. It is the starting compound for the formation of jasmonic acid, alpha-ketol, gamma-ketol and trihydroxy fatty acids. All these reactions occur during the maceration of plant tissue thus decreasing the yield of hexenals. Jasmonic acid e.g. is formed by the action of allenoxidsynthase and allenoxidcyclase and the enzymes of the ß-oxidation. Alpha-ketols and gamma-ketols are sideproducts of that reaction.

In view of the hexenal production two enzymes are important: the lipoxygenase for the formation of the hydroperoxide and the hydroperoxide-lyase. Much research has already been conducted on lipoxygenases especially those from soybean. They catalyze the addition of molecular oxygen to the molecule at carbon 13. The resulting hydroperoxide is (S)-configurated. Lipoxygenases have also been detected in microorganisms and plant lipoxygenases have been expressed in host organisms. They are available for the biotechnological production of the 13-hydroperoxide. The other decisive enzyme is the hydroperoxide-lyase. Due to the difficult isolation biochemical information is rather scarce. Recently, the construction of recombinant yeast cells containing the hydroperoxide lyase gene from banana fruit (*Musa sp.*) has been published [5].

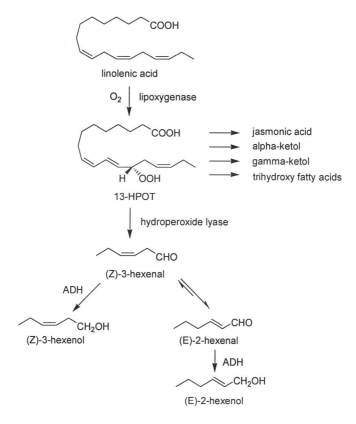

Figure 2: Formation of (Z)-3-hexenal and (E)-2-hexenal in plant tissue [4].

Researchers from Givaudan succeeded in the isolation and transfer of the lyase gene from banana to yeasts. The yeast produced hexenals from hydroperoxides. Thus the way for the microbial production of hexenals from fatty acids is open as far as the respective host has been generated containing the lipoxygenase and the lyase genes.

It is known that flax seed produces high amounts of alpha-ketols from linolenic acid via the highly unstable allenoxide. No physiological role within the plant organism has been elucidated so far for this molecule. On the other hand, many soil bacteria are known catalyzing the Baeyer-Villiger oxidation of ketones. The Baeyer-Villiger oxidation is a reaction which inserts oxygen into a keton to form an ester. Soil bacteria were screened for their ability to grow on 2-tridecanone as sole source of carbon [6]. The procedure yielded a bacteria culture tentatively identified as *Ralstonia sp.* with abundant monooxygenase activity which was used as a biological catalyst for a Baeyer-Villiger oxidation. The incubation of the alpha-ketol with the bacteria yielded (Z)-3-hexenal and (Z)-3-dodecendioic acid which can be transformed to traumatic acid the already known wound hormone (Figure 3). Thus, the enzyme system from flax seed and the monooxygenase system from the microorganism represent a promising approach for the biotransformation of linolenic acid to natural hexenals.

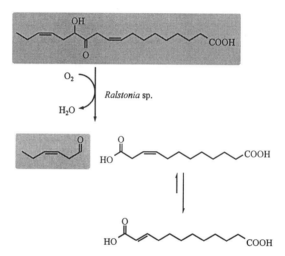

Figure 3: Formation of (Z)-3-hexenal by a biological Baeyer-Villiger-Oxidation [6].

Lactones

A second group of fatty acid derived flavor compounds are lactones or alkanolides. The naturally occurring, organoleptically important lactones generally have gamma or delta-lactone structure, and are straight-chained, while a few are even macrocyclic. Lactone flavor substances play an important role in the overall aroma presentation of many of our foods and beverages. The chain length can be even or odd numbered. However, the even numbered predominate. Lactones are important flavor substances for pineapples, apricots, strawberry, raspberry, mango, papaya, passion fruit, peach and plum. Due to their low odor threshold they have a high flavor value in the fruits. At present, lactones are made fairly expensively via chemical synthesis from keto acids. On the other hand microbially produced lactones have the advantage of being pure optically and natural. There are numerous microorganisms that are known to synthesize lactones. Lactones can be formed either by *de novo* synthesis, by ß-oxidation from ricinoleic acid, free fatty acids or hydroxy acids, by reduction from unsaturated lactones or from cheese.

The biosynthesis of lactones in plants and microorganisms is complex and not that well understood. The following systems have been implicated: i) reduction of keto acids by NAD-linked reductase, ii) hydration of unsaturated fatty acids, iii) from hydroperoxides, iv) from fatty acid epoxides, v) from naturally occurring hydroxy fatty acids, and vi) cleavage of long chain fatty acids. Most of the work on the biosynthesis of lactones has been done using microorganisms. But recently, Boland and coworker presented results on the biotransformation of gamma-dodecalactone in ripening strawberry fruits [7].

The 9,10-epoxyoctadecanoic acid, formed by epoxidation from oleic acid was proposed as the precursor for dodecano-4-lactone in strawberry fruits (Figure 4). Finally, ß-oxidation and cyclization leads to the lactone. The novel pathway relies on enzymatic reactions which are ubiquitous in plant kingdom and which are often involved in plant defense against microbial aggressors.

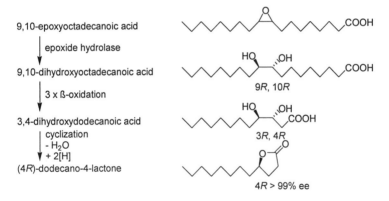

Figure 4: Biosynthesis of gamma-dodecalactone in strawberry fruits [7].

Boland and coworker performed their experiments with strawberries and nectarines and obtained a different degree of regio- and enantioselectivity for both fruit types [7]. This is in accordance with the results on the ^{13}C content of gamma-decalactones originating from various sources. In recent years the isotope ratio of ^{12}C and ^{13}C is increasingly used for the differentiation of natural, biotechnologically produced and artificial flavor. Using the highly sensitive isotope ratio mass spectrometry in connection with GC it is frequently possible to distinguish the different sources of the flavor compounds. Tab.1 presents the isotope ratio of gamma-decalactone obtained from various sources expressed as delta ^{13}C in promille referred to the standard PDB [8]. Four groups can be distinguished by their delta ^{13}C values; gamma-decalactone obtained from strawberries, stone fruits, microbial source, and synthetic source.

Table 1: ^{13}C Content of Gamma-Decalactones Originating from Various Sources [8]

Origin	delta (^{13}C) $^0/_{00}$ (PDB)
Natural	
Strawberry	-28.2 - -30.5
Peach	-38.5 - -40.9
Plum	-39.6
Apricot	-38.0
Microbial	-30.3 - -31.2
Synthetic	
Aldrich	-26.9
Roth	-24.4
Takasago	-26.0

The different values obtained for strawberries and stone fruits point to a modification of the biosynthesis of the lactones in these fruits which is in accordance with the data obtained by Boland and coworker [7].

1,3-Dioxanes

Recently, we isolated and identified a new group of volatiles, initially from apple cider and later also from apples and pears. The basic structure is that of the 1,3-dioxane (Figure 5).

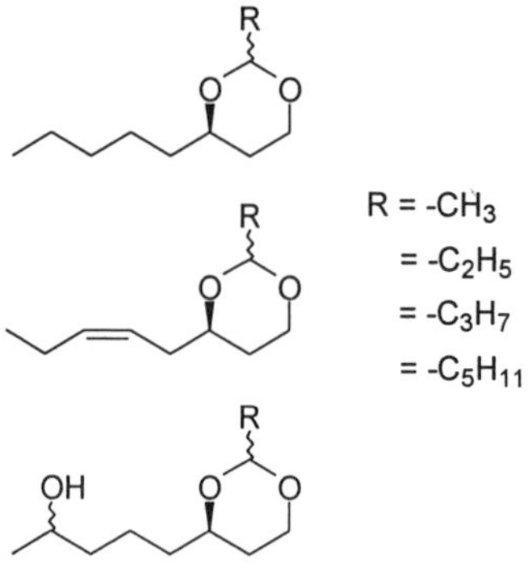

R = -CH_3
= -C_2H_5
= -C_3H_7
= -C_5H_{11}

Figure 5: Naturally occurring 1,3-dioxanes.

Generally, these compounds are formed by the reaction of aldehydes such as acetaldehyde, propanal, butanal, and hexanal with octane-1,3-diol, 5(Z)-octene-1,3-diol or octane-1,3,7-triol. The compounds posses up to three chiral centers. Carbon-3 is always (R)-configurated with an enantiomeric excess of greater than 99 %. The 1,3-dioxanes show structure analogy with 1,3-oxathianes, important volatile isolated from the yellow passionfruit.

As the 1,3-dioxanes were initially isolated from apple cider we assumed that they have been synthesized from the 1,3-diols present in apples and acetaldehyde formed by the fermentation process. However, by analyzing apples and pears we were able to identify these compounds in mature, uninjured fruits. They always appear as a diastereomeric mixture of 10:1 [9]. The odor impression of the 1,3-dioxanes have been described as green, mushroom and fruity. However, the odor thresholds have not been determined yet.

Elucidation of the chiral centers was determined as follows [10]: Starting with the enantiomerically enriched (R)- and (S)-octane-1,3-diol the respective 1,3-dioxanes were prepared by the reaction with acetaldehyde. Separation of the enantiomers of the major diastereomer was achieved on a chiral cyclodextrin column. As the natural compound coeluted with the product obtained from the reaction of (R)-octane-1,3-diol with acetaldehyde, the (4R)-configuration of the natural 1,3-dioxane was deduced. The chiral center at carbon 2 was assigned by the Nuclear Overhauser Enhancement technique (NOE). The NOE experiment showed that the proton at carbon 2 interacts with protons at carbon 4 and 7. Therefore the chair conformation was concluded leading to a (S)-configuration at carbon 2. The natural 1,3-dioxanes occur as a diastereomeric mixture of (2S, 4R) and (2R, 4R) with a ratio of 10:1.

We also investigated the biosynthesis of (R)-octane-1,3-diol and 1,3-dioxanes by using isotopically labeled substrates (Figure 6). In apples the diol is synthesized from linoleic acid and the products of the 13-hydroperoxide of linoleic acid. These compounds are primarily incorporated into the diol. Small amounts of the trihydroxy acids are also converted to the diols. However, the majority of the label originating from the trihydroxyacids is found in two still unknown compounds. The 1,3-dioxanes are formed from the aldehydes and the 1,3-diol. It is still unknown whether there is an enzyme involved in the formation of the 1,3-dioxane ring.

4-Hydroxy-3(2H)-furanones

Carbohydrates are also natural sources for flavor compounds. Although numerous investigations have been performed on the Maillard reaction few data are available on carbohydrate derived natural flavors such as 2,5-dimethyl-4-hydroxy-3(2H)-furanone (DMHF) also called furaneol® and its derivatives (Figure 7) The glucuronide is the major metabolite of the furanones after the administration of strawberries to volunteers. DMHF was identified for the first time in pineapples in 1967. In the meantime 3(2H)-furanones have been detected in numerous plants (Tab. 2). Up to now, the DMHF have only been isolated from fruits. To our knowledge there is no report on the isolation from leaves, flowers, shoots or roots. Strawberries contain high concentrations of DMHF (< 50 mg/kg).

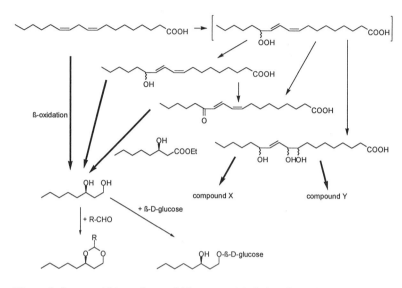

Figure 6: Proposed biosynthesis of (R)-octane-1,3-diol and 1,3-dioxanes.

In order to elucidate the biosynthesis of furanones in strawberries we applied several radioactively labeled substrates to detached ripening fruits and quantified their incorporation into the furanones [11]. On the basis of previous reports we used [2-^{14}C]dihydroxyacetone, D-[1-^{3}H]glucose, D-[U-^{14}C]glucose, D-[U-^{14}C]glucose-6-phosphate, D-[U-^{14}C]fructose and D-[U-^{14}C]fructose-1,6-biphosphate and determined the incorporation into furaneol, glycosidically bound furaneol and methoxyfuraneol. Comparison of the incorporation of the furanones showed that fructose-1,6-biphosphate is the most efficient precursor of the furanones. This observation corresponds very well with data that fructose-1,6-biphosphate is the best precursor for furaneol in *Zygosaccharomyces rouxxi* cultures [12]. On the basis of our incorporation experiments three groups of substrates were obtained. Substrates which were not incorporated into the furanones e.g. L-lactaldehyde, L-rhamnose, L-fucose, substrates with incorporation less than 0.05 % e.g. D-glucose, dihydroxyacetone, pyruvate, acetate and substrates with incorporation greater then 0.1 % such as D-glucose-6-phosphate, D-fructose, and D-fructose-1,6-biphosphate with 0.3 %. Methoxyfuraneol was formed from furaneol and S-adenosyl-L-methionine. We applied separately both radiolabeled substrates and obtained radiolabeled methoxyfuraneol.

However, there was no evidence whether D-fructose-1,6-biphosphate was directly transformed to furaneol or cleaved prior to the formation of furaneol. Therefore, we applied D-[U-^{13}C$_6$] fructose to detached ripening strawberry fruits

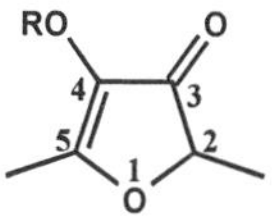

R = H   DMHF

R = CH$_3$ DMMF

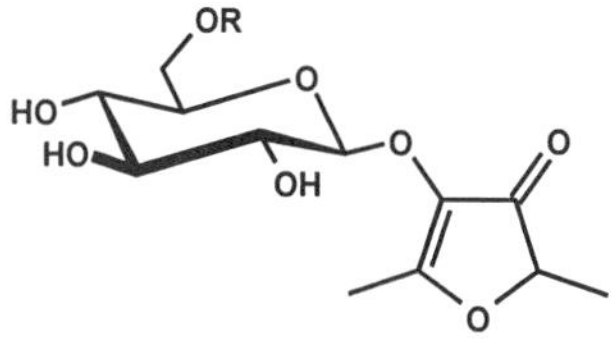

R = H   DMHF ß-D-glucopyranoside

R = COCH$_2$CO$_2$H   DMHF 6-O-malonyl ß-D-glucopyranoside

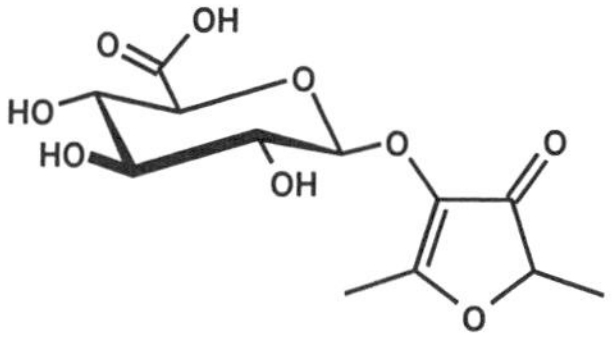

DMHF ß-D-glucuronide

Figure 7: Naturally occurring 2,5-dimethyl-4-hydroxy-3(2H)-furanone derivatives.

and we analyzed the resulting furaneol. The degree of labeling of the naturally occurring furanones, after the application of uniformly D-[U-^{13}C$_6$]fructose is presented in Tab. 3. Furaneol, methoxyfuraneol, and acetylated furaneol were all labeled by 8 %. The additional label in the glucoside and malonylated glucoside was probably located in the carbohydrate moiety. This assumption was supported by the fact that after enzymatic hydrolysis of the glucoside the aglycon furaneol showed the same degree of labeling as free furaneol.

Several biotechnological routes have been proposed for the production of natural furaneol. Most approaches have the production of 6-deoxysugar in common as 6-deoxysugars form furaneol after heating with base. One approach uses the aldolase reaction to form dihydroxyacetone which reacts with lactaldehyde in the presence of aldolase to 6-deoxyfructose-1-phosphate. The equilibrium can be shifted in favor of the product by the addition of triosephosphate-isomerase. Acid hydrolysis yields 6-deoxyfructose and during heating with base furaneol is formed. The second approach uses 6-deoxy-L-sorbose, an isomer of 6-deoxyfructose for the production of furaneol. In this case the deoxysugar is generated by the action of transketolase from 4-deoxy-L-threose and hydroxypyruvate. Hydroxypyruvate is formed from L-serine by serine-

Table 2: Occurrence of 4-Hydroxy-3(2H)-furanones in Plants

Plant Family	Species	Plant
Actinidiaceae	*Actinidia chinensis* Planch. cv Hayward	kiwi
Anacardiaceae	*Mangifera indica* Mill.	mango
Annonaceae	*Annona cherimola* L. Merr.	cherimoya
Bromeliaceae	*Ananas comosus* L. Merr.	pineapple
Cucurbitaceae	*Cucumis melo* L.	melon
Cupressaceae	*Juniperus phoenicea*	juniper
Moraceae	*Artocarpus polyphema* Pers. (Malaysia)	chempedak
Myrtaceae	*Psidium guajava* L. (Brazil)	guava
Passifloraceae	*Passiflora incarnata* L.	passion flower
Rosaceae	*Rubus arcticus* L.	arctic bramble
Rosaceae	*Rubus laciniatus* cv.Evergreen Thornless	blackberry
Rosaceae	*Fragaria*	strawberry
Rosaceae	*Rubus idaeus* L. (wild species)	raspberry
Rosaceae	*Rubus idaeus* L. x *Rubus arcticus* L.	hybrid
Rubiaceae	*Psydrax livida*	
Solanaceae	*Physalis peruviana* L.	cape gooseberry
Solanaceae	*Solanum vestissimum* D.	lulo
Solanaceae	*Lycopersicon esculentum* L.	tomato
Umbelliferae	*Levisticum officinale* Koch	lovage
Vitaceae	*Vitis sp.*	grape

glyoxylate aminotransferase. Hydroxypyruvate is also the starting material for 4-deoxy-erythrulose catalyzed by transketolase. The 4-deoxy-L-threose is generated by a microbial isomerization from 4-deoxy-erythrulose.

Table 3: Incorporation of [U-^{13}C$_6$]D-fructose into 1-5. Intensity of the Heavier Isotopomer is Expressed as Percentage of the Lighter Isotopomer

	D-fructose $^{13}C_6$	control
1: *m/z* 134/128[1] [M] $^+$ (1)	8.2 +/- 3.2	< 0.1
2: *m/z* 148/142[1] [M]$^+$ (1)	7.8 +/- 2.1	< 0.1
3: *m/z* 134/128[1,2] [M]$^+$ (1, 2)	9.3 +/- 2.5	< 0.1
4: m/z 297/291[3] [M+1]$^+$ (3)	10.9 +/- 4.2	< 0.1
5: 383/377[3] [M+1]$^+$ (3)	10.6 +/- 2.9	< 0.1
1 after hydrolysis of **4** and	8.3 +/- 0.3	< 0.1
5: *m/z* 134/128[1] [M]$^+$ (1)		

1 R = H
2 R = CH$_3$
3 R = CO-CH$_3$
4 R = ß-D-glucose
5 R = 6'-O-malonyl ß-D-glucose

(1) determined by HRGC/MS
(2) quantification of *m/z* 176/170 not feasible due to the coelution of a compound
   exhibiting the fragment ion *m/z* 176
(3) determined by HPLC/MS

Vanillin

Because of its outstanding and famous flavor quality, vanillin is highly appreciated by the consumer and most important from a commercial point of view. Vanillin is obtained from vanilla beans. Vanilla plants belong to the orchid family, existing in more than 100 species, but only two species, *Vanilla planifolia* and *V. tahitensis*, are of practical relevance. The metabolic pathway that synthesizes vanillin in the vanilla bean is still not completely understood. Nevertheless a promising start has been made in developing microbial methods for the production of vanillin. Eugenol, extracted from oil of cloves has been converted into vanillin by microorganisms such as *Arthrobacter*. Other candidates for the bioconversion by microorganisms and enzymes are coniferyl esters, curcumin and ferulic acid. These are only a few examples. The most promising transformations are those starting from inexpensive sources such as ferulic acid.

Ferulic acid is an extremely abundant and nearly ubiquitous phenolic derivative. As ferulic acid may be obtained from many sources such as sugar beet, corn cob meal, rice spelts -even a transformation of eugenol is described- it represents the most promising candidate for the biotechnological production of vanillin (Figure 8). A number of microorganisms perform the desired conversion to the aldehyde but they use the aldehyde also as carbon source. Degradation products are vanillic acid and methoxyhydroquinone. Therefore, the yields are very low. However, mutants of *Pycnoporus* degraded natural ferulic acid to vanillin without significant further conversion. Even higher concentrations of vanillin were obtained by a two-step bioconversion process of ferulic acid. In the first step, *Aspergillus niger* transformed ferulic acid to vanillic acid and in the second step vanillic acid was reduced to vanillin by *Pycnoporus cinnabarinus* [13]. Sequential addition of the precursors improved the yields. Approx. 200 mg/L vanillin were obtained with a molar yield of 22%.

It is supposed that the degradation of ferulic acid proceeds directly analogous to the well known ß-oxidation pathway of fatty acid oxidation (Figure 9). Key steps in this scheme are activation to the CoASH thioester, hydration of the enoyl-SCoA to the ß-hydroxy derivative, ß-oxidation to the ß-keto thioester, and thioclastic cleavage to give the corresponding benzoyl-SCoA together with acyl-SCoA. The aldehyde, vanillin would be formed by reduction of vanilloyl-SCoA. However, a gene encoding an enoyl-SCoA hydratase/lyase enzyme for the conversion of feruloyl-SCoA to vanillin and acetyl-SCoA was isolated from a *Pseudomonas* strain able to utilize ferulic acid as sole carbon source [14]. The gene encoded no NAD+ binding domain, and the enzyme did not exhibit ß-oxidation activity. The function was confirmed by expression in *E. coli*. This gene and its encoded enzyme offers important new possibilities for the biotechnological production of vanillin.

Isotopic analysis represent the method of choice for the determination of genuineness of vanillin (Tab. 4). The Vanilla plant belongs to the CAM (Crassulacean Acid Metabolism) plants. Therefore, vanillin extracted from vanilla beans can be distinguished from synthetic vanillin obtained from lignin, eugenol or guaiacol by ^{13}C isotope analysis. Lignin, eugenol and guaiacol are products obtained from C$_3$ plants. The C$_3$ plants use a different set of enzymes for the

84

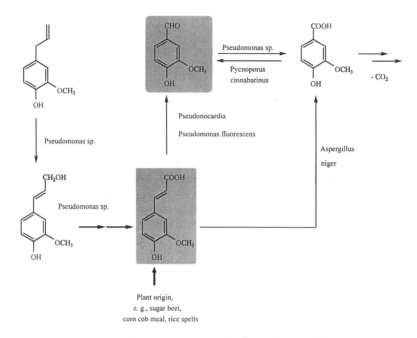

Figure 8: Bioconversion of ferulic acid to vanillin.

photosynthesis, -the fixation of carbon- than C_4 and CAM plants. Therefore, constituents of C_3, C_4 and CAM plants show different $^{12}C/^{13}C$ ratios. Vanillin isolated from *V. planifolia* has values of -18.7 to -20.8 promille and biotechnologically produced vanillin shows values of -26.9 to -32.7 promille. Blends of synthetic vanillin with isotopically enriched vanillin can be detected by site-specific isotope ratio analysis [15].

The methyl carbon of the methoxy group in the vanillin molecule can selectively be transformed to methanol by treatment with hydrogen iodide. Isotope ratio analysis of the released methanol showed a partial enrichment in ^{12}C for natural vanillin obtained from *V. planifolia* and a partial decrease of ^{12}C for vanillin isolated from *V. tahitensis*. The methyl carbon in the *V. tahitensis* samples have the isotopic composition expected from the C_4 pathway (approx. -10 promille), while all the other vanillins have the isotopic composition expected from a C_3 pathway. Since the Hatch-Slack (C_4) and Calvin (C_3) pathway are operative in CAM plants, it is possible that vanillin in *V. tahitensis* derives its methyl carbon from a C_4 pathway while vanillin from *V. planifolia* derives its methyl group from a C_3 pathway.

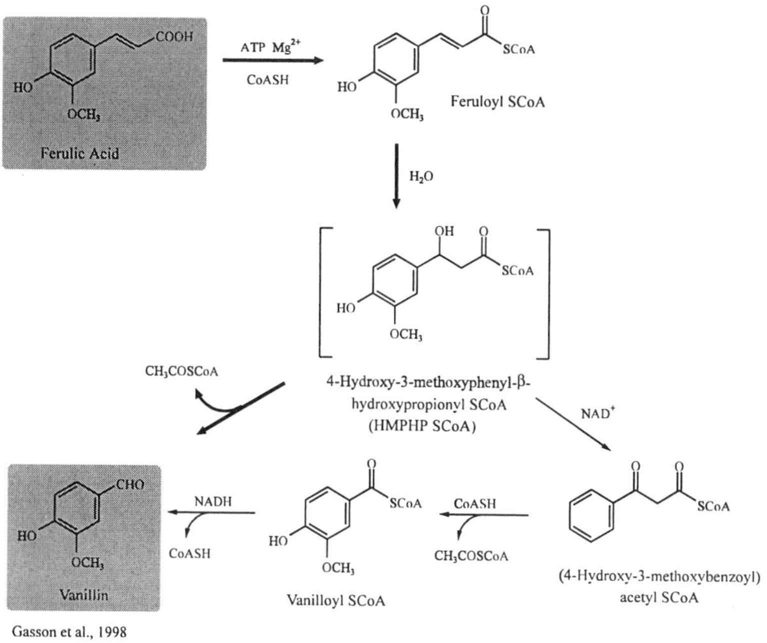

Gasson et al., 1998

Figure 9: Conversion of ferulic acid to vanillin [14].

Table 4: Isotopic Analyses of Vanillin

Origin	region	delta ^{13}C value total carbon	delta ^{13}C value methyl carbon
Vanilla planifolia	Mexico	-20.8	
	Madagascar	-20.5	-25.5
	Comoro Islands	-20.8	-25.5
	Reunion	-19.9	
	Java	-18.7	-25.3
Vanilla tahitensis	Tahiti	-16.8	-11.1
Lignin	France	-26.9	
	Switzerland	-28.2	
	USA	-27.3	
Eugenol (clove oil)		-30.8	
Guaiacol		-32.7	

References

1 Welsh, F. W. *Bioprocess Production of Flavor, Fragrance, and Color Ingredients*; Wiley: New York, 1994; p 1.

2 Manley, C. H. *Bioprocess Production of Flavor, Fragrance, and Color Ingredients*; Wiley: New York, 1994, p 19.

3 Cheetham, P. S. J. *Advances in Biochemical Engineering Biotechnology, Volume 55 Biotechnology of Aroma Compounds*; Springer: Berlin 1997, p 1.

4 Hatanaka, A.; Kajiwara, T.; Matsui, K. *Progress in Flavour Precursor Studies*; Allured Publishing Corporation: Carol Stream, 1993, p 151.

5 Häusler, A.; Schilling, B. *Flavour perception, Aroma Evaluation*; Eigenverlag Universität Potsdam: Bergholz-Rehbrücke, 1997, p 375.

6 Schneider, C.; Wein, M.; Harmsen, D.; Schreier, P. *Biochem. Biophys. Res. Comm.* **1997**, 232, 364.

7 Schöttler, M.; Boland, W. *Helv. Chim. Acta* **1996**, 79, 1488.

8 Bernreuther, A.; Koziet, J.; Brunerie, P.; Krammer, G.; Christoph, N.; Schreier, P. *Z. Lebensm. Unters. Forsch.* **1990**, 191, 299.

9 Beuerle, T.; Schwab, W. *Z. Lebensm. Unters. Forsch.* **1997**, 205, 215.

10 Dietrich, C.; Beuerle, T.; Withopf, B.; Schreier, P.; Brunerie, P.; Bicchi, C.; Schwab, W. *J. Agric. Food Chem.* **1997**, 45, 3178.

11 Roscher, R.; Bringmann, G.; Schreier, P.; Schwab, W. *J. Agric. Food Chem.* **1998**, 46, 1488.

12 Hecquet, L.; Sancelme, M.; Bolte, J.; Demuynck, C. *J. Agric. Food Chem.* **1996**, 44, 1357.

13 Lesage-Meessen, L.; Delattre, M.; Haon, M.; Thibault, J. F.; Ceccaldi, B. C.; Brunerie, P.; Asther, M. A *J. Biotechn.* **1996**, 50, 107.

14 Gasson, M. J.; Kitamura, Y.; McLauchlan, W. R.; Narbad, A.; Parr, A.J.; Parsons, E. L. H.; Payne, J.; Rhodes, M. J. C.; Walton, N. J. *J. Biol. Chem.* **1998**, 273, 4163.

15 Krueger, D. A.; Krueger, H. W. *J. Agric. Food Chem.* **1983**, 31, 1265.

Chapter 6

Natural Flavor Production Using Enzymes

Thomas A. Konar

International Bioflavors Inc., 1730 Executive Drive, Oconomowoc, WI 53066

The use of whole cell systems and isolated enzymes are discussed. Examples of hydrolysis reactons, such as the development of methylketones, are given. The discussion centers around production that can be performed using relatively basic materials.

The production of flavor compounds is generally done on a synthetic basis. This is usually the most cost effective and controlled method of production. There is however a growing perception among consumers that naturalness, and therefore natural flavors are better than synthetics. These natural flavors can thereby command a higher price than the synthetics. This not only gives an opportunity to the marketing of natural flavors but also to the chemist who creates natural flavors.

The difference between naturally and synthetically produced flavor compounds is essentially legislative. There is really no distinction organoleptically between identical compounds produced by synthetic or natural routes. Synthetics can be called nature identical if they are identical in all chemical respects to the compound as it appears in nature. If synthetics are not nature identical, they are classified as artificial. Natural flavors can be classified as WONF (with other natural flavors) or TDNS (Totally Derived From The Named Source).The focus here will be on the synthesis of natural flavor chemicals, and further, natural flavor chemicals derived from enzyme synthesis.

Any discussion that deals with the production and/or synthesis of food flavors needs to begin with the definition of flavor itself. What is a flavor? A short definition would be "that impression which is made on the taste buds when a food product is consumed." This definition does not cover any of the characteristics of the taste sensation such as the odor, bite, structure, or total physiological and physical impressions, but it is a good place to start.

As with everything, flavor and its raw materials are separated into classifications. Most of the natural raw materials can be classified into the categories in Table I.

Table I. Categories of Natural Flavor Raw Materials

A. Physical Forms	B. Compound Type
Extracts	Alcohol
Distillates	Acids
Olcoresins	Esters
Absolutes	Aldehydes
Fruit Juices	Acetals
Essential Oils	Ketones
Isolated Natural Raw Materials	Lactones
	Ethers
	Hydrocarbons
	Sulfur and Nitrogen Derivatives

There are four sources for food enzymes.

1. Enzymes endogenous to the food source
2. Enzymes from microbial contaminants
3. Enzymes from microorganisms which are desirable and deliberately added to foods
4. Isolated enzymes that are intentionally added to foods.

Unfortunately enzymes that are endogenous to a food system are generally destroyed or inactivated during processing. Enzymes that are introduced to the food system through microbial contamination generally give rise to off flavors. Microbial contamination is also contradictory to the ultimate aim of food preservation and is therefore undesirable. This leaves us with the enzymes from microorganisms that are deliberately added to food, i.e., whole cell systems and isolated enzymes intentionally added to foods. Discussion of the synthesis of the compounds in Table I will focus on the whole cell systems and isolated enzymes.

Whole cell systems are fermentations of specific microorganisms usually on inexpensive substrates. The fermentation products usually are derived from anabolic and or catabolic enzyme-catalyzed steps and are always natural. The fermentation products need to be removed from the cells or medium. This separation step is not always needed; only if the product is needed in its purified state.

Isolated enzymes or crude enzyme preparations also can produce flavor chemicals. Isolated enzymes fall into one of the six (6) categories listed in Table II.

Table II. Enzyme Categories

Enzyme	Discription
Hydrolases	This is the most important enzyme class for flavor production. Hydrolases perform hydrolysis reactions that transfer the functional groups to water. Substrates that hydrolases act on include amides, esters, epoxides, glycosides and peptides. The classes of enzymes included are amylases, esterases, hemi-cellulases, pectinases and proteases.
Isomerases	Isomerases transfer functional groups within molecules to yield isomeric forms. An example enzyme is glucose isomerase.
Ligases	Ligases form carbon-carbon, carbon-sulfur, carbon-oxygen or carbon-nitrogen bonds by condensation reactions coupled to ATP cleavage
Lyases	This group of enzymes performs the addition of functional group to carbon double bonds or the formation of double bonds by the removal of functional groups. An example is the hydroperoxide lyases.
Oxidoreductases	Oxidoreductases transfer electrons, either hydride ions or hydrogen atoms. These may require NAD(P) H as a cofactor. Examples of oxidoreductases are catalase, lipoxygenase and glucose oxidase.
Transferases	This class of enzyme transfers functional groups from one substrate to another.

The use of whole cell systems and isolated enzymes can be used in combination, separately or in tandem. In "combination" would be the use of isolated enzymes and whole cells at the same time in the same reaction vessel. In "tandem" would be the use of isolated enzymes and whole cell systems in sequential order on the very same substrate.

When isolated enzymes and whole cell systems are used separately, they can be reacted with the same type of substrate and then combined or with different types of substrates and either combined or used on their own.

The hydrolase group of enzymes and whole cell systems for flavor production is the focus of the remaining discussion. Because hydrolases are by far the most prevalent group used in industry today. For the production of flavorful materials, it is very important to keep in mind the function of the enzymes or whole cell systems as they behave in their natural state. By understanding how these active ingredients work in their "native habitat" it becomes much clearer as to how to manipulate them when they are applied in a flavor creation scheme. The development of the piquant flavor in Parmesan cheese or a tomato ripening on the vine are examples of flavors developed through enzymolysis in nature. Just as the cheese maker can produce literally hundreds of varieties of cheese types and flavors by altering the milk type, cheese making procedure, starter culture, rennet and curing of the cheese, the flavorist can manipulate similar parameters to develop enzyme-modified cheeses (EMC) that can be used as flavors. The selection of protease and lipase reacted with the cheese, coupled with the reaction parameters, give the flavorist a large palette from which to work. EMC is one of the more powerful examples of the use of enzymes and whole cell systems for this approach to flavor production. The investigation into cheese flavor has a long and varied history. While many of the volatile components of the different varieties of cheese flavors are well known, it is the middle-range compounds, such as peptides and the sulfur-containing compounds, that make EMC such an effective contributor of cheese flavor to food systems.

In EMC production enzyme selection is extremely critical. One of the major hurdles for EMC usage acceptance in industry is the bitter peptides that can develop. These are usually associated with the endoprotease activity that is a side activity of several lipases. This sometimes can be alleviated through the use of an aminopeptidase that exhibits debittering activity, but generally it is through the selection of the proper lipase that bitterness formation can be prevented.

Another example of hydrolases enzyme and whole cell system flavor creation is the development of methylketones using blue cheese mold spores for the creation of blue cheese type flavor. From an economical standpoint, the starting material is dairy fat, preferably a cream or butter selected based on the desired flavor strength and quantity of fat that would be desired in the finished flavors. A lipase of either animal or microbial source then is chosen to hydrolyze the butterfat material to free fatty acids. With the release of the free fatty acids and their subsequent availability as a substrate for the blue mold spores, the natural metabolism of the spores can convert the free fatty acids to the methyketone form. This occurs in the following manner:

$$CH_3-O-\overset{\overset{O}{\|}}{C}-R_1 \xrightarrow{\text{lipase}} HOOC-CH_2-\overset{\overset{O}{\|}}{C}-R_1 \xrightarrow{\text{mold spore}} CO_2 + CH_3-\overset{\overset{O}{\|}}{C}-R_1$$
$$\text{methylketone}$$

The only disadvantage of this method is that when made with animal derived lipase the product is not kosher. Microbial-derived lipases have been able to replace those derived from animal sources and can be used to produce these types of flavors for use in kosher products.

One of the first uses of hydrolyses in industry was the production of enzyme-modified milk high in free fatty acids. The animal lipases initially used work very well for the development of flavor materials because animal lipases (pre-gastric esterase) have a high affinity for low chain length fatty acids (C4 - C12). These are the fatty acids partially responsible for the flavor of cheese and butter. The original enzyme-modified milk produced was in the spray-dried form and it was very useful in compound chocolate coating and confectionery products. Synthesis of the free fatty acid flavor compounds is very simplistic in that the hydrolysis of the lipid material is nonspecific. While a crude preparation of the lipid hydrolysis is typically used, the preparations could be processed to derive the specific fatty acids, but this would need to be done in a very cost-effective manner on an industrial scale. The only disadvantage of using animal-derived lipase is that the product cannot be considered kosher. Microbial-derived lipases have been selected to replace those derived from animal sources and can be used to produce these types of flavors for use in kosher products.

Flavor compounds can be derived from animal or plant proteins by hydrolysis with peptidase. The initial crude preparation contains many free amino acids and large peptides that can be further broken down with the peptidase to smaller peptides and additional free amino acids. The amino acids then can be used as starting materials for further reactions to generate compounds such as phenylacetic acid and 2-methylbutyric acid. Many of the hydrolysates can be used for flavor enhancing purposes such as hydroliyzed vegetable protein, etc. and several of the free amino acids are very useful in formulations or as flavors in their own right.

Carbohydrates are good raw materials for the development of fruit-base flavors and some meat and reaction flavors. The carbohydrate is hydrolyzed using amylase or hemicellulase. The hydrolysis products can be used as is or further processed to produce specific compounds such as furanones or pyranones. They also can be combined with the peptides and amino acids from the reaction above and heated to create the brothy cooked notes that are desirable in many food flavors. One example is the conversion of fructose diphosphate to Furaneaol®.

Table III gives examples of flavor compounds that are generated by enzymatic synthesis, while Table IV gives examples of flavor compounds generated using whole cell systems.

Table III. Flavor Compounds From Enzymatic Synthesis

Flavor Material	Enzyme	Enzyme Source
Mushroom Volatiles (1-octen 3-01)	Lipoxygenases, Lyases	Agaricus Sp.
Benzaldehyde	Emulsin	Pitted Fruits
	Alcohol Oxidases/Catalase	Pichia Sp./Beef Liver
Methylanthranilate	Peroxidase	Horseradish Root
Glutamic Acid	Glutaminase	Bullera Sp.
5' Nucleotides	RNA 5' Phosphodiesterase	Microbial
	5' AMP Deaminase	Microbial
Meat Bases	Proteasese/Peptidases	Bacillus Sp.
	Peptidases	Aspergillus Sp.
Fruit Bases	Amylases	Microbial
	Hemicellulases	Microbial
Specialty Bases (e.g., Peanut, Tallow)	Protease	Microbial
	Lipases	Microbial

92

Table IV. Flavor Compounds From Whole Cell Systems

Flavor	Microorganism	Molecule
Chocolate bases	*Saccharomyces Sp.*	Isovaleraldehyde
Beverage bases	*Saccharomyces Sp.*	Ethyl esters
Fruity esters	*Geotrichum Sp.*	Ethyl-2-methyl Butyrate/ethyl tiglate
Acids	*Lactonbacillus Sp.*	Succinic Acid
Carboxyls	*Candida Sp.*	Acetaldehyde
Alcohol	*Clostridium Sp.*	Butano.ethanol
Nucleotides	*Bacillus Sp.*	Inosine
Nucleosides	*Bacillus Sp.*	5-IMP
Pyrazines	*Psuedomonas Sp.*	3 isobutyl-2-methozy-pyrazine
Terpenolids	*Streptomyces Sp.*	Cadin-4ene-1ol [(+) cadinol]

In summary, there are many avenues available to the flavor industry for the use of biochemical-derived flavors. The challenge will always be to produce these flavors in a cost-effective manner so they can compete with those flavors that are derived by synthetic means. As our knowledge of these areas grow, the biochemical avenue will become a more realistic road to use.

References:

1. Whitehead, I. 1998 Challenges to Biocatalysis from Flavor Chemistry. Food Technology.52:40.

FLAVOR CHALLENGES

Chapter 7

Flavor and Package Interactions

Sara J. Risch

Science By Design, 505 North Lake Shore Drive, Suite 3209,
Chicago, IL 60611–3427

ABSTRACT

Packaging materials are used to protect a food product during storage and
distribution.   The packaging material can help protect products from
various methods of spoilage and degradation.   Unfortunately, packaging
materials are not inert. There are several different types of reactions that
can occur, including scalping, permeation, and migration.   Scalping is the
loss of a flavor into the packaging material, and has been extensively
studied in relation to the loss of d-limonene from orange juice. Migration is
the movement of components of the packaging material into the food
product, resulting in contamination of the product and potentially an
undesirable flavor. Permeation of flavors through the packaging materials
can result in both a change in the flavor profile and a loss of flavor intensity
over time. One way that the industry has tried to counteract this dissipation
and change in flavor is by adding extra flavor. This can be a very costly
approach. New techniques have been developed to measure the flavor
barrier properties of packaging materials, and these will be discussed.

Introduction

Packaging materials are designed to provide protection to a product. In many cases,
this protection is simply a cover for the product to prevent outside contamination and
to hold the product during distribution.   For many food products, additional
protection is needed to maintain the quality of the product. The most typical types of
protection that packaging materials can provide are either moisture or oxygen
barriers. For some products, other types of gas barriers are required such as carbon
dioxide barriers for carbonated beverages.   There are standardized tests for these
properties that are published by various organizations including the American

Society for Testing and Materials (ASTM) and the Technical Association of the Paper and Pulp Industry (TAPPI) in the U.S. as well as organizations in other countries. There are a number of criteria that are used to specify packaging materials. Some of these pertain to the machinability and structural integrity of the package. The main criteria that are required to maintain the quality of the product for the desired shelf life that are typically used to specify packaging materials are the oxygen and moisture barrier properties.

When a packaging material has been properly designed, it will maintain the moisture content of a product so that it will maintain the desired texture by not either drying out or getting soggy. The oxygen barrier will maintain the desired atmosphere around the product, which is particularly important in packages that have a modified or controlled atmosphere. As an example, in the case of fried snack foods, a nitrogen flush of the package that will maintain that atmosphere and help to prevent rancidity from developing. The next major hurdle in maintaining the quality of the product is to keep the flavor profile from changing. The flavor of the product is critical in that if it does not taste good, a consumer will not purchase it again. Flavors are inherently unstable, consisting of a wide variety of different types of low molecular weight volatile organic compounds that can react with one another, react with components of the product, undergo oxidation and interact with the package itself.

Interactions

The three main types of interactions that occur between flavors and packaging materials that can result in a change in the flavor or aroma are scalping, permeation and migration. Scalping (also know as sorption) is the term used to describe the loss of one or more components of a flavor into the package itself. This is the result of those components being soluble in the packaging material. Permeation is the movement of compounds from one side of the packaging material to the other. This could be compounds in the atmosphere outside of the package permeating through to the inside of the package, resulting in contamination of the product. It could also be compounds moving from the product through the package and into the surrounding atmosphere, resulting in a loss of flavor intensity or a change in the flavor profile. The last main interaction is migration of low molecular weight components from the package itself into the product. This is of regulatory concern, as it can cause contamination of a product with unapproved indirect food additives as well as causing flavor changes in the product. In all of these cases, the main result of any of the interactions will be a change in either the flavor or aroma of the product.

Scalping of flavors

One of the first significant observations of scalping occurred with brick packs of orange juice. A number of researchers studied this phenomenon in the 1980's and early 1990's. Carlson (1) and Mannheim (2) both reported that the shelf life of 100% juice drinks was only 1 – 4 months when packaged in an aseptic package, primarily due to undesirable flavor changes. One study that evaluated the flavor changes in orange juice was carried out by Moshonas and Shaw (3). They found that the flavor of aseptically packed orange juice (250 mL flexible, multilayer cartons) was unacceptable after only one week of storage at 26 C and 2 weeks of storage at 21 C. During a six-week study, the overall flavor score decreased by 50 – 60 percent of the starting score. In addition to the sensory analysis, analytical testing by gas chromatography showed a decrease in the d-limonene concentration of approximately 40 percent and an increase in the content of α-terpineol and ethyl acetate. Not only can the sorption of flavor components by the packaging material in contact with the product cause a flavor change; it can also affect the mechanical properties of the sealant layer (4).

Another area where sorption is a concern is with recycling and reuse of plastic materials such as bottles. There are regulations in place in Europe that require plastic bottles to be reusable up to 20 times. While polyethylene terephthalate (PET) bottles, which are most commonly used for beverages, is an excellent barrier and not likely to absorb a noticeable amount of flavor from a product (5), it could be a concern when the bottle is reused. A small amount of flavor from a highly flavored product could be sorbed by the bottle and then released into a lightly flavored or unflavored product such as mineral water when the bottle is next used. The possibility for this was studied by Nielsen (6). He found that only the bottles sorbed about-1 –2 % of myrcene and d-limonene present in an orange flavored beverage after 12 weeks of storage. The bottles were then subjected to a typical industrial wash using sodium hydroxide solution. This treatment removed only about half of the materials, which had been absorbed by the PET. This indicates that it is possible to have carry over from one product to the next.

When scalping occurs, there are two main ways in which the flavor can be affected. The first is that there can be a decrease in overall intensity due to loss of the compounds in the flavor with the highest impact. It is also possible to have a change in the flavor profile. This is the result of only one or a few of the compounds being soluble while the others compounds remain in the product. When this occurs, the flavor will taste different than the original flavor, but may not necessarily be weaker.

Permeation through packaging materials

Permeation, which is the movement of compounds through packaging materials, can result in a change in the flavor of a product for two main reasons. The first is loss of flavor compounds from the product during storage. One or more components of the flavor can be lost over time if the packaging material does not provide an appropriate barrier. As with scalping of flavors by the package, permeation of flavors can result in a decrease in flavor intensity over time or a change in the flavor profile. The other change is from the contamination of the product from outside sources. This could be aromas from other products in the surrounding area such as the fragrance from soap or laundry detergent contaminating a box of crackers. Another concern that has been raised recently is the low level gasoline fumes that are often present in a convenience store that has gasoline pumps outside. It has been a concern that manufacturers have been looking at recently to insure that the packaging material has the appropriate barrier to prevent this type of contamination, which can result in undesirable flavors occurring in the product. One study that looked at the possibility of contamination from fumigation surrounding a product found that using a nylon bag provided far better protection than a polyethylene bag (7).

Migration of packaging components

Migration can cause changes in the flavor of a product by contaminating it with low molecular weight components of the packaging material. These components can include residual monomers, plasticizers, processing aids, and solvents from either printing inks or adhesives. This is another area that has been widely studied. Two books have been published that contain numerous chapters on migration and its measurement (8,9). Much of the research has dealt with the development of appropriate tests that can be used as the basis of regulations for packaging materials. In the U.S., even if a packaging material meets all of the requirements in the Code of Federal Regulations, if it causes a flavor change in the product, the product is deemed to be adulterated and subject to recall. It is important to not only insure that a package meets the legal requirements, but also that it does not impart any type of undesirable flavor in a product.

One item to note is that there are instances of off-flavors in food products that are caused by the package. These are very seldom reported in the scientific literature, instead they are the subjects of extensive internal work for a food company. When an off-flavor is detected, it is important to look at all possible sources, including the package. One such example was the development of a fruity flavor in a savory, salty snack product (10). The initial suspicion was that the flavor added to the product was contaminated. On further investigation, the fruity aroma was traced to the adhesive used in the package. The batch that had been used had levels of ethyl butyrate over 5 ppm. A revised specification for the adhesive lowered the limits for this specific compound and the problem was not seen again.

Methods of measurement

Developing appropriate testing methodologies for these interactions is a challenging task. With the other barrier properties that are specified to maintain the quality of a food product, there is only one compound or entity that must be measured in permeation tests. These include water vapor, oxygen, and carbon dioxide or other individual gases. Flavors are made up of a large number of different compounds of different chemical classes. A very simple flavor or fragrance may have 25 different compounds present while more complex flavors can have hundreds of different compounds. Many of the tests that have been used look at only one compound. The compound that has been most commonly studied is d-limonene. This was the result of the initial work on orange juice where there was dissipation in flavor intensity over time. The major component of orange oil is d-limonene although it is not the major contributor to the orange flavor. The inherent problem with this is that one compound may not be representative of the entire flavor. It is possible that the compounds may interact with each other or may change the properties of the packaging material to impact the permeability and solubility of the other compounds that are present.

Existing test procedures

There are methods of evaluation that have been reported in the literature. Hernandez et al (11) reported on various methods that had been used including both an isostatic and quasi-isostatic method. In the isostatic method, the compound(s) that is permeating through the packaging film is directly measured. In the quasi-isostatic method, the compound(s) are accumulated for a period of time and then measured. This method of accumulation will help to increase sensitivity. The detection systems varied, but one typical detector is a flame ionization detector. Another method used a mass spectrometer for analysis of the volatile constituent permeating through a packaging material (12). While this system was reported to have good sensitivity, it had the flow directly into the mass spectrometer so could analyze only one compound at a time. Also, it had the disadvantage of tying up the mass spectrometer for the entire length of the permeation test. For high barrier materials, this time could be weeks or months before the test could be completed. A procedure developed at the Fraunhofer Institute (13) uses test cells that are held in a temperature and humidity controlled room. A film is sealed into a test cell with a mixture of compounds at the desired concentration in polyethylene glycol in the base of the cell. A very slow stream of inert gas passes over the top of the film and into an organic trap such as Tenax®. The Tenax® can be extracted on a period basis, such as once a week with a series of solvents. These can be analyzed by gas chromatography and the amount of each compound permeating can be quantified. This does allow for a mixture of compounds to be tested at the same time. The biggest drawback to this procedure is the time that it takes to complete a test. Many people working in this area want much faster answers.

While various method have been published, the goal of some people in the industry is to develop an automated, standardized procedure for measuring the permeability, solubility, and diffusivity of aroma compounds through packaging materials. The methods, which have been published previously, tend to be labor intensive. There is equipment that is available today that is capable of doing automated analysis of a single component. One piece of equipment is available from Mocon and another from MAS Technologies. The basic premise of both pieces of equipment is to introduce the compound to be analyzed to one side of the packaging material in a constant concentration. The other side of the material is continually swept with an inert gas that flows into a flame ionization detector. The two main issues with both pieces of equipment is that they do not have the ability to separate compounds prior to analysis and they are not sensitive enough to measure the level of compounds that are typically present in flavors. Much higher concentrations need to be used. There is question as to whether or not using a higher concentration will be representative of what happens with the lower concentration. The equipment can be used to run relative comparisons between materials but may not give actual values for the permeation that could be expected. The testing can be effective as a screening tool. In a study reported by Mount (14), a MAS 2000 (MAS Technologies, Zumbrota, MN) was used to evaluate a number of different films. Relative barrier properties were determined for three different permeants, d-limonene, ethanol, and diacetyl. The results showed that an acrylic coating coupled with metallization provided a significantly improved barrier over just metallization or coating. This does demonstrate that films can be ranked in terms of their barrier properties to specific volatile compounds.

The other commercially available equipment is the Aromatran 1A manufactured by Mocon (Minneapolis, MN). One study using this equipment showed that relative barrier properties could be determined for four different films (15). The study went one step further and compared the results to sensory tests on products packed in those films. It was found that the two materials that had the lowest values for both permeability and solubility were able to pass a sensory test, while the two that had the highest values for permeability failed the test. The sensory test used a trained panel to determine if the products in the package were different than the standard.

Challenges

When trying to evaluate the flavor and aroma barrier properties of a packaging material, selecting a representative compound is the first challenge. There is the data available on one volatile organic compound, which could be used for comparison of different materials; however, it may not be at all representative of what may happen with the flavor of the product. It is possible to select the one main or most predominant compound in the flavor of the product in question, but again, this will only tell you what will happen with that one individual compound and not the entire flavor. The best scenario is to evaluate either the entire flavor or at least a number of the key components of the flavor.

Another challenge is the time that it takes to complete a test to determine permeability (P), solubility (S) and diffusivity (D) of a volatile organic through the packaging material can be weeks if not months. Most of the materials of interest are good to excellent barriers. It can take months if not years for equilibrium to be reached. One approach to speeding up the testing has been to use prediction. There are equations that can be used to predict P, S and D using initial data obtained during the first segment of the testing long before equilibrium has been reached. The equations hold only if the material follows Fick's first law of diffusion that states that the diffusion rate will not change with time. It is important to understand whether or not there is an interaction between the polymer and permeant that may change the polymer characteristics over time. This interaction could lead to changes in the diffusion rate and permeability of the packaging material.

The tests that have been used can give a good indication of relative properties of the materials under consideration. Further work is needed to try to set standards that can be used to test materials for their overall aroma barrier properties instead of evaluating a single component of a flavor. One possible avenue is to develop a standard mixture that could be used that would represent a variety of the different chemical compounds that can be make up a flavor. This standard could then be used to rate a packaging material in general and not require that a test be run for every different flavor that might come in contact with that particular package. As companies continue to want to have products with longer shelf lives and try to reduce the amount of packaging material being used, this type of test will become more important.

1. Carlson, V.R. *Food Technol.* 1984, *38*, 47.
2. Mannheim, C.H. *Proceedings Aspeticpak 85;* Schotland Business Research Inc.: Princeton, NJ, 1985; p. 339.
3. Moshonas, M.G.; Shaw, P.E. *J.Fd. Sci.* 1989, *54,* 82.
4. Hirose, K.; Harte, B.R.; Giacin, J.R.; Miltz, J.; Stine, C. *Food and Packaging Interactions;* American Chemical Society: Washington, DC, 1988; p 28.
5. Nielsen, T.J.; Jäagerstad, I.M.; Öste, R.E.; Wesslen, B.O. *J. Fd. Sci.*1992, 57, 490.
6. Nielsen, T.J. *J. Fd. Sci* 1994, 59, 227.
7. Scheffrahn, R.H.; Bodalbhai, L.; Su, N.Y. *J. Agric. Food Chem.* 1994, *42*, 2317.
8. *Food and Packaging Interactions;* Hotchkiss, J.H. Ed.; American Chemical Society: Washington, DC, 1988.
9. *Food and Packaging Interactions II;* Risch, S.J.; Hotchkiss, J.H. Eds.; American Chemical Society: Washington, DC, 1991.
10. Personal communication, 1990.
11. Hernandez, R.J.; Giacin, J.R.; Baner, A.L. *J. Plastic Film and Sheeting*, 1986, *2*, 187.
12. Tou, J.C.; Rulf, D.C.; DeLassus, P.T. *Anal. Chem.* 1990, *62*, 592.
13. Personal communication, 1997.
14. Mount III, E.M. *Snack Prof.* 1996, *Nov. – Dec.*, 30.
15. Risch, S.J. *Proc. TAPPI Conf,* Technical Association of the Paper and Pulp Industry, 1998.

Chapter 8

Citrus Flavor Stability

Russell Rouseff and Michael Naim

Citrus Research and Education Center, University of Florida,
700 Experiment Station Road, Lake Alfred, FL 33850

Citrus flavors are among the most desirable natural flavors and are
used in beverages, confectionery, pharmaceuticals, cosmetics, and
perfumery industries. In this chapter the stability of citrus flavors in
both oils and final products is examined. Composition and
production practices are discussed. Some of the factors which
influence citrus flavor stability such as headspace oxygen, enzymes,
impurities, packaging and elevated temperature storage are reviewed.
Techniques to stabilize citrus flavors such as encapsulation, addition
of antioxidants and removal of labile compounds are compared.
Specific decomposition pathways such as acid catalyzed hydrations
and oxidations are discussed. Finally the use of aroma units and GC-
O to determine flavor loss and off flavor formation are compared in
an example using lemon oil.

Sources, Production Practices and Composition

The major source of citrus flavors are peel oils along with the volatiles condensed from
the thermal concentration of citrus juices (essence oil and aqueous essence, sometimes
called aroma). Peel oils come from the contents of the oil gland which are found on the
fruit surface and are ruptured prior to or during juice extraction and sprayed with water
depending on equipment design (1;2). In either case the oil emulsion is first centrifuged
to separate the oil, water and small peel particles. The product stream typically contains
0.5-2% oil coming into the first centrifuge and leaves with an oil content from 70-90%.
The second stage polishing centrifuge concentrates the oil to >99%. The polished oil
still contains traces of dissolved wax derived from the peel. At temperatures above 15
-20 °C the waxes are totally soluble in the oil. However, these waxes will precipitate
at lower temperatures forming a haze in the final product. To prevent this, the oil is
usually stored at 1 °C or lower for at least 30 days to let the wax precipitate and settle
in a process called winterizing (3).

Peel oils typically contain 50-95% (+)-limonene (4). Limonene levels in citrus oils are often diminished to reduce subsequent off-flavor problems due to reactions associated with high concentrations of (+)-limonene and other terpenes. Terpene levels can be reduced by washing with 60% ethanol or by vacuum distillation in a process called folding. Distillation under reduced pressure is the more common practice. The resulting oil is higher in the oxygenated flavor rich aroma compounds and lower in (+)-limonene. Time, temperature and vacuum conditions employed in folding have a major impact on the quality of the final folded oil. Kesterson observed a 40% loss of total aldehydes at 3 fold concentration and losses increased to 50% when the oil was concentrated to 10 fold (5).

When citrus juices are evaporated to make frozen concentrate, the vapors contain not only water but most of the volatile flavoring material as well. These volatiles are recovered and separated using an essence recovery system. These systems are usually an integral part of the evaporator because the process represents an inherent component of the mass and thermal balance in the concentrating process (4). In the first stage of the juice evaporator the water is volatilized along with the aroma components. Most of the water vapor is condensed in the next evaporator effect. The low boiling aroma volatiles pass on the essence recovery system. This system consists of fractionators, chillers and condensers. After condensing, the essence forms an oil phase and an aqueous phase (2).

The aqueous phase of this condensate is called water phase essence or simply aroma and contains the polar, highly volatile "top notes", components such as low molecular weight aldehydes and alcohols (6). Commercially the product is standardized according to its alcohol content, typically 12-15% alcohol. The oil phase from the condensate is called essence oil and consists of the more non polar, low boiling terpenes, terpene alcohols, aldehydes and esters. It usually contains the fruity, sweet and green flavor compounds from the fresh juice.

The stability of citrus oils are dependent upon the matrix and environment in which they are exposed. Citrus oils may exhibit different storage stabilities depending on the composition of the volatile components present, which is, in turn, determined by the method of preparation. Cold pressed peel oils contain greater amounts of coextracted nonvolatile materials than oils prepared by solvent extraction or steam distillation (7;8) and may be responsible for their slightly improved stability. Stability within the oil glands of the peel is usually very different from that which occurs after processing and concentrating or when they are mixed in a final product. In this chapter the factors which influence flavor stability, processes to stabilize flavor and specific degradation/formation pathways will be examined. Finally the use of olfactometric techniques to assess odor quality and intensity changes and to identify which compounds are responsible for the aroma will be examined.

Composition and Structures of Citrus Volatiles

Over 200 components have been identified in citrus flavors (9). Terpenes are C_{10} compounds which comprise the largest single chemical class within citrus volatiles. Sesquiterpenes are C_{15} hydrocarbon compounds found in lower amounts in citrus volatiles. There are also terpene and aliphatic alcohols, esters, aldehydes, ketones and acids along with a small but highly significant number and quantity of hetrocyclic nitrogen and sulfur compounds. This latter group of oxygenated compounds is generally considered to produce the vast majority of the aroma impact (10;11).

Terpenes (and sesquiterpenes) can be subdivided into acyclic, cyclic and bicyclic structural categories whose general structure and typical examples are given in Figure 1. The stability of these compounds and their oxygenated analogs are related to their structure. For example, acyclic terpenes are relatively unstable compared to cyclic terpenes. The ring structure apparently adds stability to the compound which the acyclic compounds lack. Because of their relative instability and slightly aggressive aroma, acylic terpenes are not commonly used in the flavor and fragrance applications. It should be kept in mind that even though acyclic terpenes are usually drawn in the

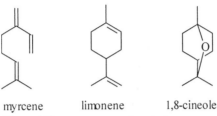

myrcene   limonene   1,8-cineole

Figure 1 Terpene structural variation

classic terpene structure shown in Fig. 1., they are in actuality compounds of generally linear structure. Myrcene, Fig. 1., is particularly reactive (unstable) because of its terminal double bond. Some of the bicyclic compounds are unstable due to one of the rings having bond angles less than the energetically favored 104°. By this measure both α and β pinene contain four membered rings which are excessively strained.

Shown in Figure 2 is the typical structural rendition for alpha-pinene on the left and a more representative structure on the right. The carbon atom with the gem dimethyl groups is shown with an asterid to make comparisons easier. Other bicyclic terpenes containing strained ring systems include sabinene and α-thujene. Alpha-pinene can undergo a number of reactions, one of the most typical is hydration with simultaneous ring opening to produce terpineol and *cis*-terpin hydrate. A more detailed discussion of the reactions of α- and β-pinene will be presented in the section on acid catalyzed reactions. Pyrolysis of α-pinene produces a mixture of ocimene and alloocimene. Commercially, α-pinene is used as the starting material for the oxidative synthesis of linalool (11).

Figure 2. Two structural representations for α-pinene.

Factors Which Contribute to Flavor Changes

Many of the early literature reports examined the combined effects of light, temperature, antioxidants and oxygen on citrus oils simultaneously (12;13). Whereas it may be useful from a practical point of view to know that light and oxygen should be excluded and antioxidants or refrigerated storage of the oils can increase stability, it is not possible to determine individual experimental factors from these reports. Another problem with the earlier literature is the incorrect characterization of many acid catalyzed hydration reactions as oxidation reactions, as clarified by Clark and Chamblee (14).

Light Exposure

In their recent evaluation of photochemical reactions involving flavor compounds, Chen and Ho (15) grouped light induced reactions into four categories depending on the presence or absence of a photosensitizer and/or oxygen. When a sensitizer and oxygen are both present, singlet oxygen can be generated that then reacts with flavor compounds containing double bonds to produce oxygenated products. Free radical mechanisms are generally involved in the other three condition categories. Thus most photochemical reactions of flavor compounds involve free radicals.

Only a limited number of studies on the influence of light on citrus oils or citrus flavors in beverages have been published. Most of these studies involve lemon or lime oils because of their commercial importance. Many of these studies suffer from experimental designs which included two or more variables changing simultaneously. There are appreciable discrepancies between the reported findings, most of which can be attributed to differences in experimental conditions or methods of analysis. The one universal finding is that citral is diminished and p-cymene is formed as a result of exposure to light. Wiley and coworkers (16) reported a turpentine-like off-odor in cola stored up to 8 weeks at 20 or 40°C under fluorescent or UV light. They suggested that the off odor was due to the formation of excess p-cymene which they found increased dramatically with increasing storage time. They also suggested that p-cymene was produced from the catalytic dehydrogenation of γ-terpinene and limonene. The fact that BHT reduced the rate of p-cymene formation suggested that free radicals were also involved. Later workers (17) examined the products of lemon oil photoxidation under UV light and an oxygen atmosphere. They reported that p-cymene did not produce a turpentine aroma. Under their GC-O conditions, p-cymene produced a solvent like odor and that a mixture of several p-menthene hydroperoxides individually and collectively was responsible for the terpentine aroma observed in stored lemon oil.

In a more recent lemon oil study (18), a serious attempt was made to separate other environmental conditions from the effects of light alone. The authors purged a lemon drink prepared from cold pressed and distilled lemon oil in 65% ethanol and 35% pH 6 aqueous buffer along with an ethanolic solution of citral in a nitrogen headspace. All samples were then exposed to UV light at ambient temperature for 4 days. The resulting GC chromatogram (carbowax column) is shown in Figure 3. Compounds 6,9 and 10 were reported for the first time. Since the authors deliberately chose to study

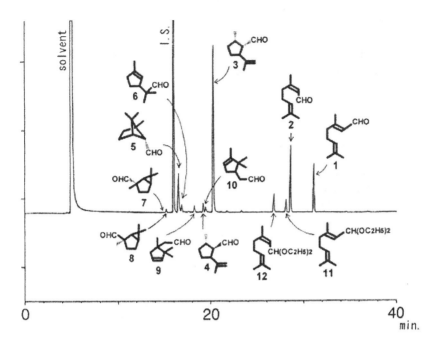

*Figure 3. Chromatogram of photoreaction products of citral in ethanol after 4 days
exposure to UV light at 30 °C. 1= geranial, 2=neral, 3=photocitral A, 4=epiphotocitral
A, 5=photocitral B, 6=2-(3-methyl-2-cyclopenten-1-yl)-2-methylpropionaldehyde,
7=trans-1,3,3-trimethylbicyclo[3.1.0]hexane-1-carboxaldehyde, 8=cis-1,3,3-
trimethylbicyclo[3.1.0]hexane-1-carboxaldehyde, 9=(1,2,2-trimethyl-3-cyclopenten-1-
yl)acetaldehyde, 10=α-campholenealdehyde, 11=geranial diethyl acetal, 12=neral
diethyul acetal. IS = internal standard = 2-octanol (50 µg) from (18).*

these reactions under mildly acid conditions (pH 6) to minimize the competition from
acid catalyzed hydration reactions, it is not known if these same products would be
formed under more typical high acid (pH 2) conditions.

Elevated Temperature

Citrus flavor components will decompose at different rates depending primarily
temperature and pH. Decomposition reaction rates generally follow the Ahrenius rate
relationship, which indicates that the rate of the reaction doubles for each 10 °C
increase in temperature or conversely, decreases for each 10°C decrease. Thus, most
citrus flavors are stored at reduced temperatures to minimize decomposition reactions.

Storage Studies

In a three month study involving orange juice packaged both in glass and Tetra-Pak laminated soft containers (19), there was a significant loss in limonene due to absorption into the polyethylene package liner. The concentration of α-terpineol, a reputed off-flavor formed from limonene, increased more rapidly at higher storage temperatures. Storage temperature rather than initial limonene concentration had the greater effect on α-terpineol concentrations. Interestingly, the rate of increase was greater in glass bottles than in soft packages stored at the same temperature. The juice was described as stale and musty after 13 days at 32 °C, or 90 days at 20 °C (62 days in glass), but was still acceptable after 3 months at 4 °C.

Accelerated Storage Studies

Flavor instability of citrus can sometimes takes weeks or months before sensory differences are detected. Shelf life studies are designed to determine how long a product is viable under specific storage conditions. Both color and/or homogeneity/ viscosity will degrade with increased storage, however, the predominant factor in determining shelf life for most products is usually flavor deterioration. This is especially true for citrus juices. Early investigators (20-22) were quick to discover that juice flavor could be maintained for extended periods at low temperature (1-4°C) storage but was degraded more rapidly at higher storage temperatures. The obvious temptation was to store samples at increasingly higher temperatures to more rapidly determine what might occur for longer storage periods at lower temperatures. It has been recently reported (23) that orange juice quality changes during storage for up to half a year may be predicted by monitoring concentrations of selected components during 1-2 weeks accelerated storage at 50 °C. Other investigators (24) examined orange juice thermal degradation reactions by heating juice to 75, 85 and 95°C for 0, 15, 30 and 60 minutes. As shown in Figure 4, the formation of 4-vinylguaiacol is highly temperature dependent as noted in earlier studies carried out at lower temperatures (25-27). However, the normal commercial practice would be to heat orange juice to 95-98 °C for only a few seconds (28) to inactivate enzymes. Thus it would be difficult to extrapolate these observations to more typical situations without first examining the same system at lower temperatures and longer times. It should also be kept in mind that different reactions will have different temperature dependancies. Thus an off flavor reaction that takes place rapidly at highly elevated temperatures, will become the major flavor reaction at elevated temperatures. However, it may only be a minor reaction at much lower temperatures. A hypothetical example is shown in Figure 5. In this example 2,3-dimethylpyrazine would be the dominate flavor compound formed at 160°C, but would be a relatively minor component at 30 °C. Conversely, HMF would be the dominate compound formed at 30°C, but a minor component at 160°C. Thus, the relative composition of aroma volatiles formed at high temperature *may* be very different than what is formed at lower temperatures. Therefore, extrapolation of findings at high temperatures and short times to predict what may occur at lower temperatures at longer times should be done with caution.

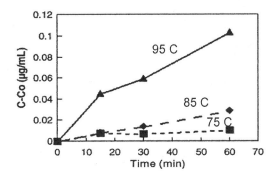

Figure 4. Temperature dependence of 4-vinylguaiacol formation in orange juice at elevated temperature. Where C = measured concentration, C_o = initial concentration, from (24).

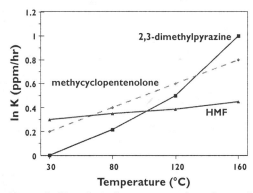

Figure 5. Hypothetical temperature dependence of for the relative rates of formation of three flavor compounds, adapted from (29).

Headspace/Dissolved Oxygen

Haro Guzman (30) investigated the effects of atmospheric oxygen and ambient light on the stability of distilled lime oil. As shown in Figure 6, significant losses in γ-terpinene, terpinolene and α-terpinene and almost a four fold increase in p-cymene were

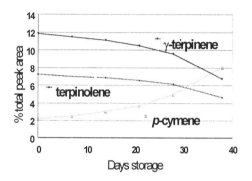

Figure 6. Ambient temperature storage of distilled lime oil exposed to air, adapted from (30).

observed during 38 days storage.

Even though Iwanami and coworkers (18) purged their lemon flavor solution with nitrogen, they also found limonene oxides after 4 days ambient storage under UV radiation. These limonene oxides were thought to be due to the reaction of limonene and residual dissolved oxygen. Shown in Figure 7 are the chromatograms of the control sample stored in the dark and the identical lemon flavor sample exposed to UV light for 4 days at 30°C.

Peaks designated with asterisks (*) were limonene oxides. However, if the limonene oxides were due to the reaction of limonene and oxygen alone, then these same oxide peaks should be present in the control as well. Since the control chromatogram has little if any limonene oxide peaks, the exposure to light is apparently necessary to produce limonene oxides.

Enzymes

These protein based catalysts can produce profound chemical changes in many food systems. Juices are particularly prone to enzyme mediated changes as the extraction process disrupts the cells where enzymes have been compartmentalized. Many of these enzymes played significant roles in developing flavor components during various stages of fruit maturity. However, after the fruit is macerated to liberate the juice, only a

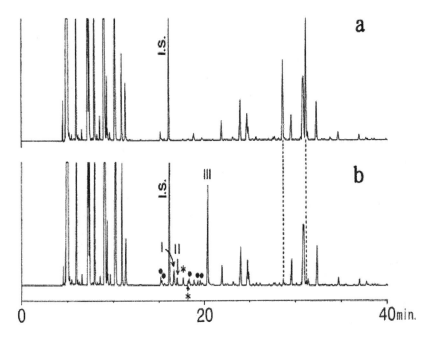

Figure 7. Chromatogram of lemon flavor stored under two conditions; (a) stored at 30 °C in dark, (b) stored under UV irradiation at 30 °C , IS = internal standard, 2-octanol, from (18).

limited number of flavor altering enzyme systems are active. At ambient temperature (25°C), orange juice esterase activity juice declines after approximately 2 hours, while phosphatase activity maintains relatively constant (31). Since the concentrations of fatty acids in citrus juices are so low (32), any flavor compounds formed from enzymatic cleavage of these compounds would a have marginal overall contribution to observed flavor changes. The major enzymatic reactions in freshly squeezed citrus juices involve pectin methyesterases. Many studies have reported physical changes associated with pectin methyesterases, but no reports have been found which pectinase activity directly altered the flavor of the final product.

Native enzymes have little if any activity in citrus oils or essence products. Since enzymes are proteins of relatively high molecular weight, they would not be sufficiently volatile to be distilled and condensed with aqueous essence or essence oil. Cold pressed oil would probably not contain active enzymes because any enzymes would likely be denatured from the high (50-95%) limonene (terpene) content.

Packaging

Container characteristics can have a profound influence on the flavor stability of the product (33). Packaging contributes to flavor changes in one of three basic modes.

First the packaging material might act as a source of flavor contaminantion which leaches into the product during storage ultimately changing the flavor profile of the product. Volatile low molecular weight chemicals are often added to plastics (polymers) to improve their functional properties. Examples include: plasticizers to improve flexibility, UV "blockers" to prevent discoloration and antioxidants to prevent oxidation of the plastic. Other packaging based flavor sources include: unintentional manufacturing contaminants such as polymerization accelerators, cross-linking agents, antistatic chemicals, lubricants, etc., (34;35).

The second mode of flavor change is through the absorption of flavor materials from the product into the package (typically polymers). The absorption can be nonspecific so that there is a general decrease in flavor intensity or it can be selective, where only certain flavor components are preferentially absorbed. This latter case will produce flavor imbalances in the product. Duerr (19) reported that (+)-limonene was preferentially absorbed into polyethylene-lined cartons (40% in six days), but desirable flavor compounds were only marginally absorbed. These findings were confirmed in later studies (36;37). Duerr and others (38) suggest that the absorption of limonene was an advantage, in that limonene was not a major flavor impact compound but was the starting material for a significant storage off-flavor, α-terpineol. Thus the loss of limonene would not diminish flavor and might reduce the potential for subsequent off-flavor formation. Later studies report similar findings with respect to the loss of limonene and minimal change in flavor (39;36).

Whereas there appears to be substantial agreement that limonene is significantly adsorbed by polymer packaging, the literature is less consistent with respect to flavor impact compounds absorbed. Shimoda and coworkers (40) reported distribution ratios (film:juice) were 1.2-1.7, 0.65, and 0.19-0.24 for terpene hydrocarbons, terpene aldehydes and terpene alcohols, respectively after 7 days storage. The latter two groups contain compounds considered to have significant flavor impact. Thus approximately 35% of the terpene aldehydes and approximately 80% of the terpene alcohols were adsorbed. One possible source for this apparent discrepancy is due to the different polymers studied. Earlier studies examined low density polyethylene whereas the latter study examined polyethylene terephthalate (PET). Multilayer laminated packaging material can offer some flavor stability improvements. It has been reported (41) that important flavor aldehydes from orange juice were absorbed by "juice board" (paper board sandwiched between two layers of low density polyethylene) but were absorbed to a smaller extent and at a slower rate by "barrier board" (paper board sandwiched between two layers of low density polyethylene with an inner layer of ethylene vinyl alcohol copolymer).

The third mode in which packaging can influence flavor is where packaging materials allow external factors such as light or oxygen to interact either directly with the flavors in the product or with product components to produce additional flavors. These factors are discussed separately in other sections of this chapter.

Contaminating Impurities

Using packed column GC with 2 different stationary phases, Bielig and coworkers (42) reported that at pH 3.5, in the presence of Fe or Sn, valencene can be oxidized by atmospheric oxygen to nootkatone, a characteristic flavor component of grapefruit.

They suggested that this reaction was responsible for the development of a bitter, grapefruit-like flavor which was observed in canned (tin coated steel) orange juice during storage but not detected in the same juice packed in glass and stored under similar conditions.

Flavor Stabilizing Techniques

The stability of citrus flavoring materials and juices can be improved through low temperature storage and the exclusion of oxygen. Other techniques include:

Encapsulation

Citrus oils can be encapsulated with a variety of water soluble materials which act as oxygen barriers of differing efficiencies and offers the additional advantage of presenting the flavoring material as a pourable powder. An excellent overview of encapsulated of spray-dried flavors, including citrus, was presented by Brenner (43). The relative advantages and disadvantages of various matrix materials, stability to oxidation and volatilization, recovery of flavor oils, emulsion formula and drying conditions, and economics are examined.

The stability of encapsulated orange peel oil using maltodextrins of DE values of 4, 10, 20, 25 and 36.5 as encapsulating agents was evaluated by Anandaraman and Reineccius (44). Samples were stored at 32, 45 and 60°C, and examined periodically using high resolution capillary GC The GC profile and sensory quality of the stored product was compared with control encapsulated samples stored at 4 ° C. The oil displayed distinct signs of deterioration at elevated storage temperature as evidenced by increased levels of limonene-1,2-epoxide and carvone (oxidation products of (+)-limonene). Less deterioration was observed with maltodextrins of higher DE values suggesting the possibility that these materials posse superior O_2-barrier properties. It was suggested that by increasing the DE by 10, a 3-6 fold improvement of shelf life could potentially be achieved.

Bhandari and coworkers (45) investigated the microencapsulation of lemon oil using ß-cyclodextrin. using a precipitation method at the five lemon oil to ß-cyclodextrin ratios of 3:97, 6:94, 9:91, 12:88, and 15:85 (w/w) in order to determine the effect of the ratio of lemon oil to ß-cyclodextrin on the inclusion efficiency of ß-cyclodextrin for encapsulating oil volatiles. The retention of lemon oil volatiles reached a maximum at the lemon oil to ß-cyclodextrin ratio of 6:94; however, the maximum inclusion capacity of ß-cyclodextrin and a maximum powder recovery were achieved at the ratio of 12:88, in which the ß-cyclodextrin complex contained 9.68% (w/w) lemon oil. The profile and proportion of selected flavor compounds in the ß-cyclodextrin complex and the starting lemon oil were not significantly different.

Kopelman et al., (46) developed a freeze drying method for the production of water soluble citrus aroma powders to be used as natural flavour ingredients in soft drink dry mixes. They reported a retention of approximately 75% of the initial aroma volatiles using the optimal maltodextrin 15 DE/sucrose (3:2) carrier. This would be a remarkable achievement given the extremely high volatility of the components in citrus aroma such as acetaldehyde, methanol and ethanol.

Antioxidants

It has been known for some time that citrus flavors can be stabilized to a degree through the use of antioxidants (13). There have also been a few reports of extracts from juice or oil possessing the ability to inhibit the oxidation of limonene (47;48).

Removal of Labile Components

Since γ-terpinene was thought to be one of the most unstable monoterpenes which contributed little to the aroma but whose decomposition products produced off flavors, Ikeda and coworkers (49) proposed the selective removal of γ-terpinene from lemon oil. The partially concentrated lemon oil would have an odor similar to that of natural oil but with improved stability. However, the relative stability of γ-terpinene is not entirely clear. Verzera and coworkers (50) reported that γ-terpinene was among the most stable monterpenes stored in aqueous citric acid solution (pH 2) during for 3 months at 25 °C. It could not be determined if these model solutions were exposed to light or oxygen. Haro Guzman (30) reported a 44% loss of γ-terpinene in undiluted lime oil during a 38 day ambient temperature study in which the oil was not protected from the light or oxygen (See Figure 6). Unfortunately, no internal standard was employed it could not be determined if some γ-terpinene was lost through evaporation. In evaluating these conflicting reports it should be pointed out that the work of Verzera and coworkers (50) was carried out in dilute citric acid solution similar to commercial soft drink conditions whereas the work of Haro Guzman (30) was done with the undiluted oil. Differences between these two reports may be two the result of considering two types of stability, that is, stability of the raw material (oil) vs stability of the oil in a final product.

In addition to significant losses of γ-terpinene, Haro Guzman (30) also reported losses of 68% and 36% for α-terpinene and terpinolene respectively along with a 270% increase in p-cymene. He hypothesized that if the more unstable compounds were removed the resulting oil should have greater stability. Through selective fractional distillation he was able to produce a 4 fold oil that had reduced levels of (+)-limonene, γ-terpinene and terpinolene. This oil along with a standard 4 fold lime oil was also stored for 38 days. p-Cymene was measured (as area per cent) and used as a measure of instability. The reduced terpene oil contained less than half the p-cymene (compared to the control) at the end of storage period. The reduced terpene oil was reported to impart a fresher more complete odor as compared to the standard 4 fold oil (control). Unfortunately sensory details were not provided, so it is not known if this was a single self evaluation or the results of a blind study by a sensory panel.

Citrus Flavor Degradation/Formation Pathways

Terpene Oxidations

Because citrus flavors are usually found in aqueous acidic environments, the major mechanism by which oxygen is added to unsaturated terpenes is through acid catalyzed hydration and not oxidation (14). Never-the-less oxidation reactions do occur, particularly in other environments. Oxidation of orange and lemon oils typically produce limonene peroxides and carvone (47;51;17;38). Oxygen will attack either or both the endocyclic (1,2) and exocyclic (7,8) double bonds forming a mixture of limonene hydroperoxides such limonene-2-hydroperoxide (2-hydroperoxy-p-mentha-6,8-diene) among others. Using high resolution capillary gas chromatography, Schieberle and coworkers (52) were able to resolve six limonene hydroperoxides from the photooxidation of (R)-(+)-limonene. (Rose Bengal was used as an oxidative catalyst). Limonene peroxides have been reported to contribute to the "turpentine" off aroma often noted in heavily oxidized orange oil (17). Since carvone is one of the major decomposition products of limonene, it has been proposed to be used as an indication of citrus oil oxidation (13;44). Interestingly some minor hydroperoxides of d-limonene were reported to be potent contact (skin) allergens in guinea-pigs (53).

Some of the cyclization products of citral, namely, p-mentha-1,5-dien-8-ol and p-mentha-1(7),2-dien-8-ol, are oxidized to p-cymene-8-ol unless oxygen is vigorously excluded (54). This alcohol is dehydrated to form p-α-dimethylstyrene (14), one of the final products of citral cyclization. In an oxygen environment as much as 80% of the end products ends up as p-α-dimethylstyrene whereas in a nitrogen environment only 17% of the final products ends up as p-α-dimethylstyrene and 81% as p-cymene (55).

Citral Decomposition

Citral is an extremely important flavor component of citrus oils, especially lemon and lime oils where it may constitute at least 50% of the oxygenated fraction of the oil. It is responsible for the fresh lemony/ citrus aroma so highly prized in many products. In an aqueous model system at ~pH 3, a 15 ppm solution of citral slowly lost its lemony flavor and developed a bland, mild fruity taste after partial deaeration and ambient temperature storage in the dark. When the identical experiment was carried out at ~75 ppm citral, an oxidized, terpen, objectionable taste was observed after storage (14).

Citral actually consists of two geometric isomers, neral and geranial, generally in the ration of 2:3. These isomers are stable enough to be isolated in high purity (>90%) at least for a short period of time. Eventually each will revert back to the mixture of the two forms. The cyclization decomposition reaction of citral is fairly complex due to a number of secondary oxidation and dehydration reactions which are very condition dependent. A thorough discussion of this topic is beyond the scope of this work. A simplified reaction scheme is shown in Figure 8. The final products of citral decomposition are p-cymene and p-α-dimethylstyrene. As previously discussed, the relative distribution of these two products is highly dependent on the presence of oxygen. As shown in Figure 8, both neral and geranial undergo proton attack to form a common oxonyium intermediate (A) which converts to intermediate B. This intermediate can undergo a minor side reaction which will not be discussed here. As long as citral has not been depleted, then the major products of the reaction are the alcohols D, F, G and H. Along with the loss of desirable flavor due to the cyclization of citral, is the production of "turpentine" and other oxidized flavors. The compounds responsible for these flavors have been determined using gas chromatography-olfactometry, GC-O, and will be discussed in one of the following sections.

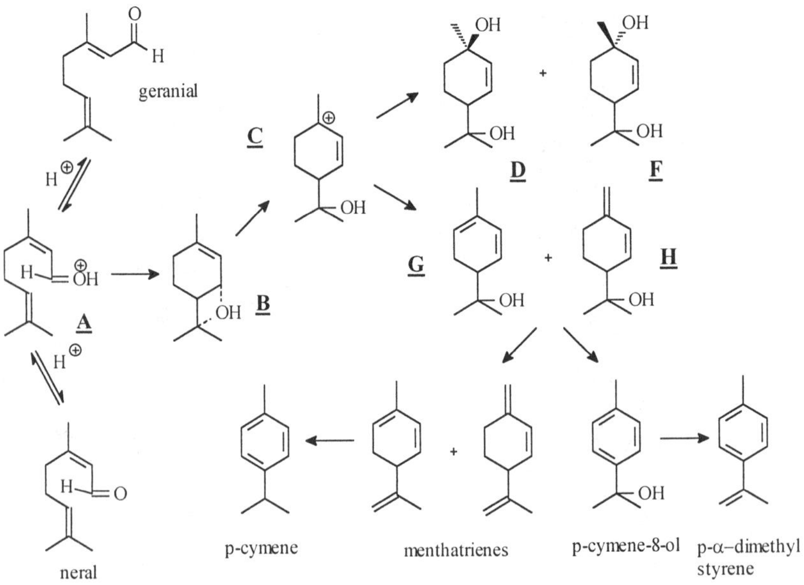

Figure 8. Decomposition reaction pathway for the cyclization of citral in aqueous acid environment. Adapted from (14).

Acid Catalyzed Hydrations

These reactions involve the electrophilic addition of water across a double bond in an acid environment. Citrus flavors are often found in acidic conditions as this is their natural environment. Terpenes, are usually olefinic (hydrocarbons containing double bonds) and terpenes comprise the largest single chemical class in citrus flavors. Thus acid catalyzed hydrations are a major reaction mechanism for citrus flavors.

The electrophilic addition of water across a terpene double bond usually involves two steps. The first step (the rate determining step) involves the electrophilic attack of a hydrogen (hydronium) ion on the terpene double bond, forming a carbonium ion intermediate. The second step (rapid) involves the reaction between the positive carbonium ion and a negative species, in this case OH⁻ to form an alcohol. This reaction follows Markovnikov's rule (the hydrogen goes to the carbon atom which has the greatest number of hydrogens). The hydration of limonene (A) can be used as an illustration of this reaction. In the upper sequence, the exocyclic double bond is attached by the acid to ultimately form α-terpineol (C). Terpinolene (not shown) is also formed from carbonium ion B. In the lower sequence the endocyclic double bond is attacked to ultimately form *cis* and *trans* ß-terpineol (E, F). In actuality the reaction is somewhat more complex with additional side reactions and rearrangements. It should also be

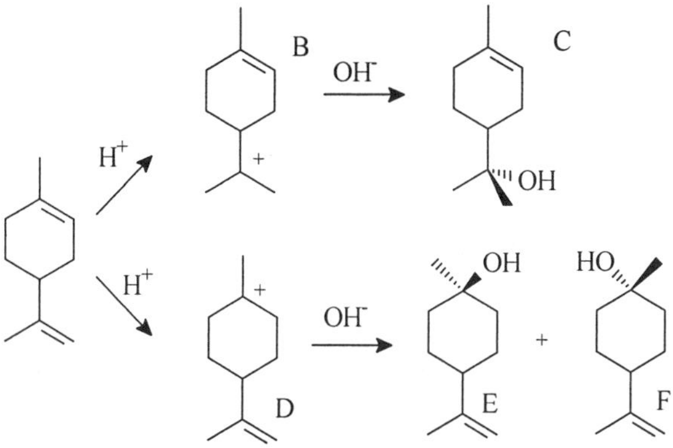

Figure 9 Generalized acid catalyzed hydration reactions of (+)-limonene (A).

pointed out that each of the terpineols formed above (C,E,F) all contain an additional double bond which can also undergo an electrophilic addition of water to form *trans*- and *cis*-1,8 terpin. Additional details can be found in the excellent discussion by Clark and Chamblee (14) or in the more general review by Ohloff and coworkers (56).

Since there are many terpenes in citrus which contain double bonds, the logical question is which terpenes react more rapidly (are most unstable). Fortunately certain general rules can be applied. Exocyclic double bonds are more readily attacked than endocyclic double bonds and conjugated double bonds are the slowest to react. In the example with limonene, α-terpineol is formed 10x faster than *cis* and *trans* ß-terpineol (14).

Verzera and coworkers (50) examined the relative stabilities of five monoterpenes (myrcene, α-terpinene, γ-terpinene, limonene and terpinolene) in aqueous citric acid solution (pH 2) during storage for 3 months at 25 °C. γ-Terpinene and myrcene were stable under these conditions. Limonene and α-terpinene lost about 25% of their original concentration during storage, but almost 85% of terpinolene had decomposed. In reality there are always competing reactions of breakdown products. In addition to acid catalyzed hydrations, isomerization and oxidations reactions also take place. Additional information on terpene stability can be found in the section entitled Removal of Labile Compounds.

Both cyclic and acyclic sesquiterpenes can also undergo acid catalyzed reactions. For example, the farnesenes can form the corresponding alcohols, and be oxidized to the corresponding aldehydes (sinensals), with tremendous change in sensory properties.

Nootkatone, can undergo acid catalyzed hydration of its exocyclic double bond to form a keto alcohol as shown below. Nootkatone is a sesquiterpene ketone which possesses grapefruit like aroma character and a bitter taste. It is an important flavor impact compound in grapefruit flavors and most grapefruit oil is currently sold on the basis of its nookatone content. At pH 2.4, half the nootkatone can be converted to the much less valuable keto alcohol shown above in about three weeks.

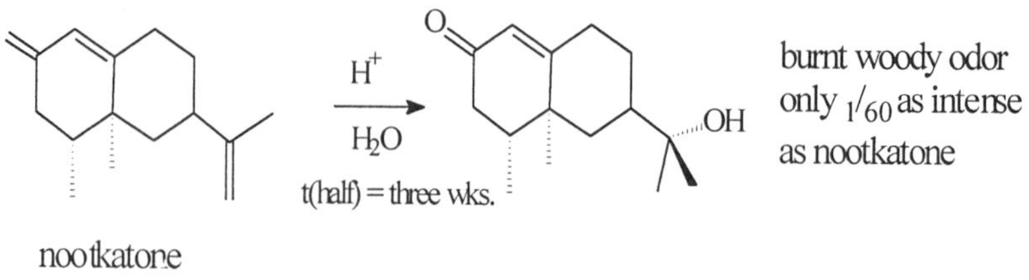

Figure 10. Acid catalyzed hydration of nootkatone.

GC-Olfactometry Identification of Flavor Changes

One of the major considerations in evaluating flavor stability is to determine exactly which components have aroma activity. Not all volatiles in a sample extract will have the same relative aroma activity as observed in the original sample, as some components will be minimized and others concentrated in the sample preparation process. Aroma activity is usually calculated by dividing the concentration observed in the sample by the aroma threshold for that compound. The resulting ratio is called the aroma value. If the ratio is greater than one the component should have aroma activity. If the ratio is less than one, no aroma activity should be observed. The value of this approach is that all volatiles can now be compared on an equal basis (i.e., their aroma strength) regardless of differences in concentration or aroma potency. There are two limitations to this approach. First, it requires an accurate aroma threshold value be available for the component in the sample matrix. In practice this is rarely available. Most published aroma threshold values are given for a water matrix. Thus, an aroma value calculated using a water threshold can only be considered approximate because threshold values in food samples are usually higher than those in water. Secondly, it requires that component concentration values reflect the concentration in the original sample and not the extract. Many investigators fail to consider extraction efficiencies and simply employ extract concentrations.

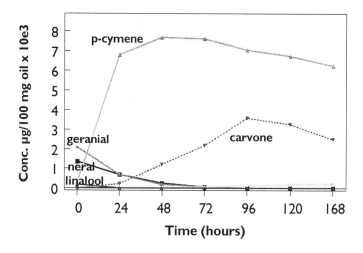

Figure 11. Quantitative analysis of aroma components in lemon oil during storage. Adapted from (57).

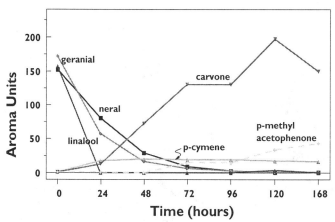

Figure 12 Peroxidized lemon oil changes measured in aroma units. Adapted from (57).

The analytical data from Schieberle and Grosch's studies on the peroxidation of lemon oil has been plotted for discussion purposes. They first measured concentrations of the various reactants and products in oxidized lemon oil as show in Figure 11. It should be noted that the compound present in highest concentration is p-cymene. Other compounds such as p-methylacetophenone are also plotted but are lost in the baseline because their concentrations are so small. However, if their aroma values are calculated using published aroma thresholds then the relative importance of carvone and p-methylacetophenone can be seen. Even though p-cymene was actually present in highest concentration, its aroma threshold was also high. Its probable contribution to the observed storage off flavors is relatively small. Thus the use of aroma units allows the determination of which components will impart the greatest aroma impact.

Another useful approach to the study of off flavors has been to employ a human assessor to directly measure which components in a gas chromatographic effluent have aroma activity. Two major approaches are used: dilution analysis o r time-intensity. A comparative discussion of these two approaches is beyond the scope of this

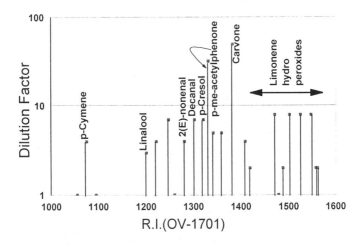

Figure 13 Lemon oil volatiles from oxygenated storage under UV light for 120 h., from (17).

presentation. Schieberle and Grosch employed aroma dilution analysis, AEDA, to the study of lemon oil oxidation and flavor changes. In this manner, measured not calculated aroma strengths are determined. Shown in Figure 13 are the relative aroma strengths (expressed as dilution values) of the products and remaining reactants of lemon oil exposed to light and oxygen at 120 h. In comparing the relative aroma strengths from Figure 12 at 120 h to those in Figure 13, it should be kept in mind that the vertical axis in Figure 12 is linear whereas the same axis in Figure 13 is logarithmic. As expected carvone has the most aroma impact. However, the major advantage in using gas chromatography - olfactometry, GC-O, is the ability of detect the presence of unexpected aroma impact components. One does not need to know ahead of time which components to measure. Using this technique many highly potent aroma impact compounds are detected that would have been missed simply because they were present at such low analytical concentrations. As an example, the peaks that elute after carvone were newly identified hydroperoxides which were responsible for the terpentine off flavor. The ability of GC-O to detect unidentified aroma impact components is one of the major strengths of this approach. The GC-O approach does have its limitations, however, as it can not determine synergistic or antagonistic interactions from other aroma active components or components in the sample. However, it is an excellent technique to determine off-flavor components in citrus oils.

Acknowledgment

The author would gratefully acknowledge the financial support of BARD, The United States - Israel Binational Agricultural Research and Development Fund, project No: US-2914-97.

Literature Cited

1. Ting, S. V.; Rouseff, R. L. *Citrus fruits and their products*; Marcel Dekker, Inc.: New York, N.Y., 1986; Vol. 18.
2. Kimball, D. A. *Citrus Processing: Quality Control and Technology*; Van Nostrand Reinhold: New York, 1991.
3. Anon *The orange book*; Tetra Pak Processing Systems, AB: Lund, Sweden, 1997.
4. Redd, J. B.; Hendrix, C. M. In *Fruit Juice Processing Technology*; 63-109, Ed.; Agscience, Inc.: Aurburdale, 1993; pp 713.
5. Braddock, R. J. In *Citrus Nutrition and Quality*; S. Nagy and J. A. Attaway, Eds.; American Chemical Society: Washington, D.C., 1980; pp 273-290.
6. Johnson, J. D.; Vora, J. D. *Food Technology* **1983**, *37*, 92-93.
7. Shaw, P. E. *J. Agric. Food Chem.* **1979**, *27*, 246-257.
8. Lund, E. D.; Shaw, P. E.; Kirkland, C. L. *J. Agric. Food Chem.* **1981**, *29*, 490-494.
9. Maarse, H.; Visscher, C. A. *Volatile Compounds in Food - Quantitative Data,*; TNO-CIVO Food Analysis Institute: Zeist, The Netherlands, 1985; Vol. 4.
10. Ohloff, G. *Scent and Fragrances*; Springer-Verlag: Berlin, 1994.
11. Bauer, K.; Barbe, D.; Surburg, H. *Common Fragrance and Flavor Materials*, Third ed.; Wiley-VCH: Weinheim, 1997.
12. Mannheim, C. H.; Passy, N. *pp* **1972**, 39-63.
13. Garnero, J.; Roustan, J. *Rivista Italiana Essenze, Profumi, Piante Officinali, Aromi, Saponi, Cosmetici, Aerosol* **1979**, *61*, 203-209.
14. Clark, B. C., Jr.; Chamblee, T. S. *Dev Food Sci. Amsterdam : Elsevier Scientific Publications* **1992**, *28*, 229-285.
15. Chen, C. W.; Ho, C. T. In *Process-Induced Chemical Changes in Food*; , 1998; pp 341-355.
16. Wiley, R. C.; Louie, M. K.; Sheu, M. J. *Journal of Food Science* **1984**, *49*, 485-488, 497.
17. Schieberle, P.; Grosch, W. *Zeitschrift fuer Lebensmittel Untersuchung und Forschung* **1989**, *189*, 26-31.
18. Iwanami, Y.; Tateba, H.; Kodama, N.; Kishino, K. *J. Agric. Food Chem.* **1997**, *45*, 463-466.
19. Duerr, P.; Schobinger, U.; Waldvogel, R. *Alimenta* **1981**, *20*, 91-93.

20. Rymal, K. S.; Wolford, R. W.; Ahmed, E. M.; Dennison, R. A. *Food Technology* **1968**, *22*, 1592-1595.

21. Askar, A.; Bielig, H. J.; Treptow, H. *Deutsche Lebensmittel Rundschau* **1973**, *69*, 360-364.

22. Koch, J. *Fluessiges Obst* **1973**, *40*, 42-48.

23. Petersen, M. A.; Tonder, D.; Poll, L. *Food Quality and Preference* **1998**, *9*, 43-51.

24. Marcotte, M.; Stewart, B.; Fustier, P. *Journal of Agricultural and Food Chemistry* **1998**, *46*, 1991-1996.

25. Tatum, J. H.; Nagy, S.; Berry, R. E. *Journal of Food Science* **1975**, *40*, 707-709.

26. Peleg, H.; Naim, M.; Zehavi, U.; Rouseff, R. L.; Nagy, S. *Journal of Agricultural and Food Chemistry* **1992**, *40*, 764-767.

27. Naim, M.; Schutz, O.; Zehavi, U.; Rouseff, R. L.; HalevaToledo, E. *Journal of Agricultural and Food Chemistry* **1997**, *45*, 1861-1867.

28. Moshonas, M. G.; Shaw, P. E. *Journal of Food Quality* **1997**, *20*, 31-40.

29. Reineccius, G. *Source Book of Flavors*, 2nd ed.; Chapman and Hall: New York, 1994.

30. Haro Guzman, L. In *Flavors and fragrances: a world perspective*; B. M. Lawrence; B. D. Mookherjee and B. J. Willis, Eds.; Elsevier Science Publishers BV: Amsterdam, Netherlands, 1988; pp 325-332.

31. Bruemmer, J. H.; Roe, B. *Proceedings of the Florida State Horticultural Society* **1975**, *88*, 300-303.

32. Nordby, H. E.; Nagy, S. *Journal of Agricultural and Food Chemistry* **1979**, *27*, 15-19.

33. Goldenberg, N.; Matheson, H. R. *Chem. Industry* **1975**, *5*, 551.

34. Kim, H.; Gilbert, S. G.; Hartman, T. In *Frontiers of flavor*; G. Charalambous, Ed.; Elsevier Science Publishers BV.: Amsterdam, Netherlands, 1988; pp 249-257.

35. Kim-Kang, H. *Crit. Rev. Rood Sci.* **1990**, *29*, 255.

36. Pieper, G.; Borgudd, L.; Ackermann, P.; Fellers, P. *Journal of Food Science* **1992**, *57*, 1408-1411.

37. Sadler, G.; Parish, M.; Davis, J.; Vanclief, D. In *Fruit Flavors: biogenesis, characterization and authentication*; R. L. Rouseff and M. M. Leahy, Eds.; American Chemical Society: Washington, D.C., 1995; pp 202-210.

38. Kutty, V.; Braddock, R. J.; Sadler, G. D. *Journal of Food Science* **1994**, *59*, 402-405.

39. Ohtsu, K.; Hashimoto, N.; Innoue, K.; Miyaki, S. *Brewers' Digest* **1986**, *61*, 18-21.

40. Shimoda, M.; Nitanda, T.; Kadota, N.; Ohta, H.; Suetsuna, K.; Osajima, Y. *Journal of Japanese Society of Food Science and Technology [Nippon Shokuhin Kogyo Gakkaishi]* **1984**, *31*, 697-703.

41. Marsili, R. In *Techniques for Analyzing Food Aroma*; R. Marsili, Ed.; Marcel Dekker: New York, 1997; pp 237-264.

42. Bielig, H. J.; Askar, A.; Treptow, H. *Deutsche Lebensmittel Rundschau* **1972**, *68*, 173-175.

43. Brenner, J. *Perfumer & Flavorist* **1983**, *8*, 40-44.

44. Anandaraman, S.; Reineccius, G. A. *Food Technology* **1986**, *40*, 88-93.

45. Bhandari, B. R.; D'Arcy, B. R.; Bich, L. L. T. *Journal of Agricultural and Food Chemistry* **1998**, *46*, 1494-1499.

46. Kopelman, I. J.; Meydav, S.; Wilmersdorf, P. *Journal of Food Technology* **1977**, *12*, 65-72.
47. Ina, K.; Hirano, K. *Journal of Food Science and Technology [Nihon Shokuhin Kogyo Gakkai shi]* **1973**, *20*, 567-571.
48. Ifuku, Y.; Maeda, H. *Journal of Japanese Society of Food Science and Technology [Nippon Shokuhin Kogyo Gakkaishi]* **1978**, *25*, 687-690.
49. Ikeda, R. M.; Stanley, W. L.; Nannier, S. H.; Rolle, C. A. In *Chem. Abs. 66, 49216b (1966)*; , 1966.
50. Verzera, A.; Duce, R. d.; Stagno D'Alcontres, I.; Trozzi, A.; Daghetta, A. *Industrie delle Bevande* **1992**, *21*, 217-222.
51. Wilson, C. W.; Shaw, P. E. *Journal of Agricultural and Food Chemistry* **1975**, *23*, 636-638.
52. Schieberle, P.; Maier, W.; Grosch, W. *Journal of High Resolution Chromatography and Chromatography Communications* **1987**, *10*, 588-593.
53. Karlberg, A. T.; Shao, L. P.; Nilsson, U.; Gafvert, E.; Nilsson, J. L. G. *Archives of Dermatological Research* **1994**, *286*, 97-103.
54. Clark, B. C.; Powell, C. C.; Radford, T. *Tetrahedron* **1977**, *33*, 2187-2191.
55. Baines, D. A.; Jones, R. A.; Webb, T. C.; Champion-Smith, I. H. *Tetrahedron* **1970**, *26*, 4901-4913.
56. Ohloff, G.; Flament, I.; Pickenhagen, W. *Food Rev. Int.* **1985**, *1*, 99-148.
57. Schieberle, P.; Grosch, W. Lebensmittel-Wissenschaft&Technologie, **1988**, 21, 158-162.

FLAVOR ANALYSIS

Chapter 9

Issues in Gas Chromatography–Olfactometry Methodologies

Jane E. Friedrich and Terry E. Acree

Department of Food Science and Technology, Cornell University,
New York State Agricultural Experiment Station, Geneva, NY 14456

In order to discriminate the small fraction of odor-active volatiles from the much larger number of odorless volatiles present
in food, most flavor chemists use gas chromatography - olfactometry (GC/O). Using GC/O to assign odor activity to components of mixtures, extracts and distillates of food and fragrance materials allows flavor chemists to focus on the components that may contribute to flavor. There are several GC/O
techniques in use but all combine olfactometry, the use of humans to access odor activity, with the gas chromatographic
(GC) separation of volatiles. In its simplest form GC/O or
"GC-sniffing" is used to determine which chemicals in mixtures
have odor so that they can be characterized using standard
chemical tools such as mass spectroscopy, infrared spectroscopy, organic synthesis, etc. Once determined they are selectively quantified in order to explain sensory properties. The
processes of predicting or explaining the sensory perceptions of
a mixture in terms of its chemical composition has had limited
success. This paper will outline the issues involved in the application of GC/O to measurements of food flavor.

GC/O In The Context Of Flavor Analysis

The study of flavor chemistry can be approached via two different methods and as
such employs two groups of scientists; analytical chemists and sensory scientists.
The technique used to study flavor differs between these two groups. Analytical chemists approach the study of flavor chemistry by measuring all the volatile chemicals
present where as sensory scientists attempt to correlate sensory data with analytical
data. However, the sensory scientist knows that most of the analytical data does not
measure the cause of the sensory responses. The technique of gas chromatography -
olfactometry (GC/O) brings these two groups of scientists together because it provides
sensory responses to chromatographically separated chemicals. The sensory scientist

finds the human responses useful and convincing, while the analytical chemist finds the retention times at which the responses were made *(1)* useful for chemical identification.

Gas chromatography - mass spectrometry (GC/MS) is a powerful tool for the separation and characterization of chemicals whether they are odor-active or not. In the analysis of flavor, GC/MS can selectively focus on the odor-active compounds once their spectral and chromatographic properties are known. However, the task of determining which compounds in a sample are odor active requires a bioassay. In other words, we must first determine "which constituent or constituents is/are contributing to the characteristic sensory properties of the food product being investigated" *(2)*. GC/O is a bioassay that reveals odorants in terms of their pattern of smell-activity thus eliminating odorless compounds from consideration. As shown in Figure 1 the most odor-active compound in peppermint oil, as determined by GC/O, was not detectable in the GC/FID chromatogram while menthol, the major volatile in the oil, contributed little to the GC/O chromatogram. Although methanol is not an odor-active compound of peppermint oil it is important to the sensory character of peppermint oil i.e. peppermint oil is not recognized as such without menthol. As GC/O analysis illuminates the most potent odorants in a mixture, traditional methods of organic chemistry (extraction, distillation, concentration, chromatography, spectroscopy, etc.) are still required to isolate and identify the relevant unknowns.

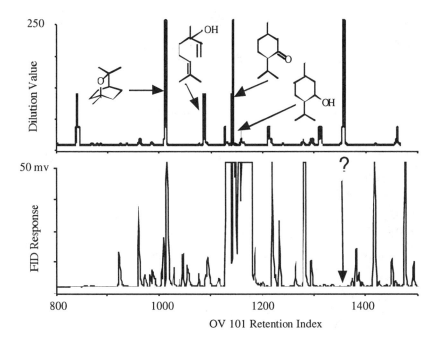

Figure 1. GC/O and GC/FID of natural peppermint oil.

Gas chromatography - olfactometry (GC/O) is the collection of techniques that combines olfactometry with the gas chromatographic (GC) separation of volatiles in order to assess odor activity in defined air streams. The first published account of GC/O analysis i.e. the direct sniffing of the effluent from a gas chromatographic column is over 35 years old (3). In the 1970's, GC/O effluents were combined with humid air under laminar flow conditions to make the process of sniffing more ergonomic and to take advantage of the high resolution in narrow bore open tubular columns (4). The use of quantitative dilution methods to assess odor potency was developed simultaneously in both the US (5) and Germany (6) in the mid 1980's. In its simplest form a GC/O is any GC modified to allow sniffing of the effluent. Because GC/O directly exposes humans to chemicals safety concerns are of paramount importance, as discussed previously by Acree (1).

GC/O analysis has some drawbacks, many of which are directly related to the use of a human as a detector. GC/O is time intensive and typically only uses 1-2 sniffers, who must be pre-screened for sensitivity and specific anosmia. It has been shown that the olfactory sensitivity of an individual changes throughout the day as well as over longer periods of time (7) (8). This is of concern because dilution analysis often takes a number of weeks to perform. To reduce experimental error and variation the following variables can be closely monitored: sample preparation, room and sample temperature, time of day, duration of analysis, repetition of analysis, repeated standardization of sniffers, and use of a standard lexicon. It is well documented that people can be trained to consistently identify smells if they are standardized periodically and trained to sniff with standard chemicals and vocabularies (9). Recent studies have substantiated the odor-relevancy of the results from dilution-to-threshold GC/O by demonstrating the similarity of a synthetic solution based on GC/O data to the food product itself (10) (11).

What Can We Measure And What Will It Mean?

Sampling Bias

Methods of sampling produce bias in data that effects the usefulness of that data for predicting properties. For example, GC/O coffee data collected by three different procedures, solid phase microextraction (SPME) headspace analysis, solvent extraction with freon 113, and stable isotope dilution assay, was found to produce different patterns of GC/O data, Table 1. The sampling bias caused by the different sample preparation methods is large. The most potent odorant in coffee was found to be dependent on the sampling method used; each sampling method resulted in a different aroma profile.

The most common methods to isolate unknown odorants from food products are distillation and solvent extraction combined with chromatography. However, introduction of artifacts, loss of highly volatile components during concentration, and co-elution of odorants with the solvent can distort the results.

Table 1. Sampling Bias Shown by the Variation of GC/O Data of Coffee Sampled by Three Different Methods (from the data of *(12)*).

	Odor Spectrum Values (i)		
Compound	SPME extraction	Solvent extraction	Stable isotope dilution assay (ii)
2,4,5-trimethylthiazole	100	33	0
sotolon	34	100	37
2-furfurylthiol	14	90	100
3-methoxy-2-isobutyl pyrazine	54	42	32
abhexon	31	25	0.2
vanillin	29	64	0.7
furaneol	26	57	49
4-vinyl guaiacol	24	70	20
4-ethyl guaiacol	15	19	0.2
2-isopropyl-3-methoxy pyrazine	15	10	0

(i) determined using Steven's Law exponent of 0.5.
(ii) from the data of *(13)*.

A method to isolate flavor compounds that is less prone to bias is headspace analysis. It is superior to extraction in that non-volatile artifact-forming compounds are not extracted therefore the results are more representative of the system sampled. There are two types of headspace analysis: static and dynamic headspace analysis. Static headspace analysis involves sampling air equilibrated above a food sample followed by direct injection into a gas chromatography - mass spectrometer (GC/MS) for identification and quantitation. Compounds that would normally co-elute with the solvent can be detected by static headspace analysis. However, the level at which the volatile compounds are present restricts this analysis. To be detected by GC/MS analysis volatiles must be present at levels equal to or greater than 10^{-5} g/L. The concentration of volatiles above a food product range from 10^{-11} to 10^{-4} g/L *(14)* and humans can smell some aroma compounds in food that have concentrations less than 10^{-12} g/L *(15)*, therefore only the most abundant volatiles will be detected by this method.

Dynamic headspace analysis uses a gas to purge the volatiles from a sample and is often more sensitive than static headspace methods *(14)*. Purge-and-trap analysis is a dynamic headspace method that traps the volatiles purged from a sample on an absorbent solid. The trap is designed to concentrate the volatiles, which are subsequently desorbed for GC/MS analysis. An advantage of this method is the ability to concentrate the product by a hundred-to a thousand-fold *(16)*.

The development and use of solid phase microextraction (SPME) as an alternative to the traditional methods has grown enormously in recent years *(17)*. This technique is highly sensitive, inexpensive, portable, reproducible and rapid, and is a solvent-free method for extraction. The overall procedure involves the direct exposure of a SPME fiber to the headspace of or immersed in a sample in a closed vial. Once equilibrium has been reached the fiber is introduced into a GC/MS for separation, identification and quantification. This is an exciting development because SPME is "amenable to automation and can be used with any gas chromatograph or mass spectrometer" *(18)*.

The application of solid phase microextraction (SPME) for the preparation of samples for GC/O dilution analysis has been shown to be feasible *(12)* although the available absorbents bias' the results in a manner similar to solvent extraction. This method could be invaluable for the analysis of heterogeneous systems, fatty foods, solid foods and aqueous foods where non-volatiles create artifacts during solvent extraction.

Combining SPME technology with retronasal aroma simulation, such as the RAS, would allow researchers to determine the aroma experience at the olfactory epithelium. Two methods have been effective in releasing the odorants with a composition that simulates the vapors at the olfactory epithelium: retronasal aroma simulation (RAS) *(19)* and gas chromatography olfactometry headspace (GCO-H) *(11)*. The composition of these headspace samples can then be analyzed chemically and combined with odor threshold values transforming them into odor activity. Alternatively a bioassay can be applied directly to the headspace to determine the activity of each constituent.

Distillation is the oldest sampling method used. Steam distillation and high vacuum stripping of oils *(14)* employ harsh conditions (high temperature and pressure) resulting in the formation of artifacts or loss of some components. In contrast, solvent extraction is more quantitative usually separating almost all of the flavor compounds from food *(14)*. The environmental hazard created by organic solvents is becoming a serious challenge of extraction methods. Other drawbacks include the time-consuming procedures, loss of sample during isolation and the need to concentrate the sample before analysis.

Overall, if a qualitative approach to flavor analysis is desired then the dynamic headspace and SPME sampling techniques have definite advantages related to their ability to concentrate the sample volatiles thus facilitating structure identification. However, any quantitative aspects of both these methods are directly related to the sampling time and thus must be monitored closely *(20)*. All methods of sample preparation produce bias in the data that must be considered when applying the results.

Diversity of Sniffers

Two distinctly different goals define most GC/O studies. One is based on the desire to use a single person *of defined properties* as a bioassay probe to study differences in odorants and the other is the use of a single odorant *of defined properties* to study differences in people. In the first case the objective of the research is to extrapolate the results to a population of samples where as the objective in the second case is to extrapolate the results to a population of subjects.

In some cases a single sample is tested using a single sniffer, for example, the most potent odorants in one sample of Rambutan fruit is sniffed by one person to determine what needs to be identified chemically *(21)*. It is in a sense a *case study* and is usually the initial phase in the analysis of a flavor system. The case study identifies the variables that should be measured in subsequent sample surveys, clinical trials or consumer testing. However, GC/O case studies are often associated with within-individual irreproducibility as shown by Abbott *et al. (22)*. The precision of threshold detection of an individual makes it more difficult for the sniffer to detect the end of an odor experience than the beginning of the experience. Figure 2 shows the large differences in the standard deviation (from Table 2) of individual responses at the beginning and end of a GC peak.

Table 2. Precision Associated with Threshold Detection Found in GC/O Analysis (from the data of (22)).

| Sniffer | Standard deviation | |
	Front of peak	Back of peak
A	3	8
B	1	2
C	1	4

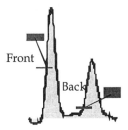

Front

Back

Figure 2. Standard deviation associated with threshold testing.

In other applications GC/O has been used to perform what may be called *clinical trials* in which the responses of different people to an odorant are studied. In one study, GC/O revealed differences in the response of subjects to six odor standards even though they were selected to be of normal sensory acuity (23). Similarly, Abbott *et al.* (23) used GC/O to determine the response of six sniffers to a beer standard and found that the number of odor active regions detected ranged from 5 to 14. Table 3 shows the between-individual variation associated with these GC/O clinical trials.

Both case studies and clinical trials raise the issue of how to standardize human subjects for GC/O experiments. In a case study we want to reduce the within-individual irreproducibility and verify that the individual sniffer has the appropriate acuity. Clinical trials add further complexity of between-individual variation to the already present within-individual irreproducibility. GC/O, like many other sensory protocols, is affected by the difference between individuals' ability to smell (23).

Table 3. Number of Odor-Active Regions Detected by Each Panelist in a Beer Sample.

Sniffer	Number of odor-active regions detected in a beer sample
A	5
B	7
C	7
D	13
E	13
F	14

130

Measurement

The data produced by GC/O has a qualitative component in which the sniffers describe the nature of their perception. This usually involves the association of the precept with a word or group of words in a lexicon. Quantitative GC/O data are either measurements of odor potency or perceived intensity plotted against retention index. Estimates of perceived intensity may also be plotted against retention index to produce OSME plots. Flavor dilution (FD) plots are measurements of potency or multiples of threshold while OSME plots are psychophysical perceptions of odor strength.

Two methods to estimate potency in GC/O effluents are aroma extract dilution analysis (AEDA) and CharmAnalysis™. AEDA is a bar graph of maximum FD values while CharmAnalysis™ yields a plot of FD values assembled objectively from the raw data. The data in both cases expresses the relative number of dilutions until a sniffer no longer detects an odor as it elutes from the column. Under ideal conditions an AEDA bar graph is identical to a bar graph made from the peak maximum in a charm chromatogram. However, charm values differ from the corresponding AEDA FD values because charm values are based on peak areas where as AEDA FD values are based on peak heights. These two methods of analysis are widely used therefore it is imperative that there is a way to compare the results.

Converting data from different types of potency measurements can be accomplished using the idea of an odor spectrum. An odor spectrum is potency data normalized to a single odorant. An odor spectrum can be further enhanced by transforming it to reflect the compressibility of olfaction, a process that also yields data with normalized error rates *(24)*. Both flavor dilution values and Charm values can also be converted to comparable odor spectrum values (OSV) using Steven's law:

$$\Psi = k\Phi^n$$

Where Ψ is equal to the perceived intensity of a stimulant, k is a constant, Φ equals the stimulus level, and n is Steven's exponent. Steven's law exponent for odorants range between 0.3 and 0.8 *(25) (26)*, using a median value of 0.5 is adequate. Odor spectrum values are usually normalized to the most intense odorant. The odor spectrum represents a pattern of the odorants in the sample independent of concentration. Another way to represent potency data is through odor activity values (OAVs). An odor activity value (OAV) is the ratio of the concentration of an odorant to its odor threshold determined in the food matrix. Under ideal conditions OAVs are proportional to the FD or Charm values *(21) (27)* especially when the dilution analysis is performed on a simulated retronasal headspace. OAVs can also be transformed into OSVs so they can be compared directly to OSV data from GC/O analysis as shown in Table 1. If the samples are prepared properly they will represent the pattern at the olfactory receptor *(28)*.

Standardization

There is not a one-to-one correspondence between a stimulant, an olfactory receptor, and a precept. For example, in rats it has been shown that increased expression of a single gene leads to greater sensitivity to a small subset of odorants, not to a single compound *(29)*. Thus, the receptor sites of mammals seem to be broadly tuned across different chemicals to create a specific aroma class. If humans have a limited set of receptors, a standard set can be designed to test all receptor sites and odor categories. Future research will test the hypothesis that a standard set of chemicals based on aroma category is a better determinant of acuity than standard solutions based on chemical class. However, the central problem of GC/O, standardization of individual sniffers, must be addressed.

A standardization procedure based on aroma categories could reduce bias and measure specific anosmia precisely. Bias can be minimized using a standard set of odorants to test the sensitivity of a sniffer and to this end a set of chemicals to standardize sniffers for GC/O analysis is currently being developed *(30)*. The standard set of chemicals consists of a flavor genus of 26 different odors drawn from a list of 23 smell categories for food aroma *(31)* plus three additional non-food aroma categories. The new genera are: maillard, dairy, edible oil, fermented, fishy, shellfish, berry, citrus, pome, stone, tropical, grain, cured meat, fabricated meat, processed meat, raw meat, herbs, peppers, roots spices, seed spices, sweet spices, aromatic, vegetable, floral, animal, and mineral *(32)*. Using this method it is hoped that a standard set of chemicals of between 50 and 100 compounds can be generated. They will be grouped into two standard mixtures. Using these standard sets of chemicals individuals can be screened for specific anosmia yielding coefficients of response for each category. We can then eliminate sniffers with general anosmia and characterize the specific anosmia of those used. Publishing measurements of the specific anosmia of sniffers used in GC/O experiments would allow us to compare data from different sniffers and different laboratories eliminating the need to use average data from multiple sniffers *(30)*. Because the size of the panel needed to extrapolate GC/O data to a larger population is prohibitive standardization is a practical and meaningful alternative.

Literature Cited

1. Acree, T.E. In *Flavor Measurement*; Ho, C.T. & Manley, C.H., Eds.; Marcel Dekker, Inc.: New York, NY, 1993; 77-94.
2. Teranishi, R. In *Flavor Analysis. Developments in Isolation and Characterization*; Mussinan, C.J. & Morello, M.J., Eds.; ACS Symposium Series 705; American Chemical Society: Washington D. C., 1998; 1-6.
3. Fuller, G.H., Steltenkamp, G.A. & Tisserand, G.A. *Annals. N.Y. Acad. Sci.* **1964**, 116, 711-724.
4. Acree, T.E., Butts, R.M., Nelson, R.R. & et al. *Anal. Chem.* **1976**, 48, 1821-1822.
5. Acree, T.E., Barnard, J. & Cunningham, D.G. *Food Chem.* **1984**, 14, 273-286.
6. Ullrich, F. & Grosch, W. *Z. Lebensm. Unters. Forsch.* **1987**, 184, 277-282.
7. Köster, E.P. *Int. Rhin.* **1965**, 1, 57.
8. Köster, E.P. *Olfactologia* **1968**, 1, 43.
9. Cain, W.S. *Science* **1979**, 203, 467-470.
10. Grosch, W., Preininger, M., Warmke, R. & Belitz, H.-D. In *Aroma: Perception, Formation, Evaluation*; Rothe, M. & Kruse, H.P., Eds.; Deutsches Institut for Ernahrungsforschung: Potsdam-Rehbrucke, 1995; 425-439.
11. Guth, H. & Grosch, W. *J. Agric. Food Chem.* **1994,** 42, 2862-2866.

12. Deibler, K.D., Acree, T.E. & Lavin, E.H. *J. Agric. Food Chem.* **1999**, *unpublished*.
13. Semmelroch, P. & Grosch, W. *J. Agric. Food Chem.* **1996**, 44, 537-543.
14. Reineccius, G.A. In *Source Book of Flavors*; Reineccius, G.A., Ed.; Chapman and Hall: New York, NY, 1994; 24-60.
15. Fazzalari, F.A. *Compilation of Odor and Taste Threshold Values Data*; American Society of Testing and Materials: Philadelphia, 1978.
16. Teranishi, R. & Kint, S. In *Flavor Science: Sensible Principles and Techniques*; Acree, T.E. & Teranishi, R., Eds.; American Chemical Society: Washington D. C., 1993; 137-167.
17. Arthur, C.L. & Pawliszyn, J. *Anal. Chem.* **1990**, 62, 2145-2148.
18. Arthur, C.L., Potter, D.W., Bucholz, K.D., Motlagh, S. & Pawliszyn, J. *LC/GC, the magazine of separation science* **1992**, 10, 656-661.
19. Roberts, D.D. & Acree, T.E. *Journal of Agricultural and Food Chemistry* **1995**, 43, 2179-2186.
20. Coleman, W.M. & Lawerence, B.M. *Flavour and Fragrance Journal* **1997**, 12, 1-8.
21. Ong, P.K.C., Acree, T.E. & H., L.E. *J. Agric. Food Chem.* **1998**, 46, 611-615.
22. Abbott, N., Etievant, P., Issanchou, S. & Dominique, L. *J. Agric. Food Chem.* **1993**, 41, 1698-1703.
23. Marin, A.B., Acree, T.E. & Barnard, J. *Chemical Senses* **1988**, 13, 435-444.
24. Acree, T.E. & Barnard, J. In *Trends in Flavour Research*; Maarse, H. & Van Der Heij, D.G., Eds.; Elsevier Science B.V., Noordwijkerhout, Netherlands, 1994; 211-220.
25. Stevens, S.S. *Science* **1958**, 127, 383-389.
26. Stevens, S.S. *Am. Sci.* **1960**, 48, 226-253.
27. Acree, T.E. *Analytical Chemistry* **1997**, 69, 170A-175A.
28. Friedrich, J.E. & Acree, T.E. *Int. Dairy Journal* **1998**, 8, 235-241.
29. Zhao, H., *et al. Science* **1998**, 279, 237-242.
30. Friedrich, J.E. & Acree, T.E. *Standardization of human subjects for gas chromatography – olfactometry (GC/O)*; XIII ECRO Congress Abstracts: Siena, Italy, 1998; 118.
31. Civille, G.V. & Lyon, B.G. *Aroma and flavor lexicon for sensory evaluation: terms, definitions, references, and examples*; American Society of Testing and Materials: West Conshohocken, 1996.
32. *Flavornet. Gas chromatography – olfactometry (GCO) of natural products*, URL http://www.nysaes.cornell.edu/fst/faculty/acree/flavornet.

Chapter 10

Studies on Potent Aroma Compounds Generated in Maillard-Type Reactions Using the Odor–Activity–Value Concept

Peter Schieberle[1], Thomas Hofmann[2], and Petra Münch[2]

[1]Institut für Lebensmittelchemie der Technischen, Universität München,
Lichtenbergstrasse 4, D–85748 Garching, Germany
[2]Deutsche Forschungsanstalt für Lebensmittelchemie, Lichtenbergstrasse 4,
D–85748 Garching, Germany

The Maillard reaction between free amino acids and reducing carbohydrates is an important tool used in the industrial production of reaction or processed flavors. However, although certain potent odorants are known to be formed from distinct precursor amino acids, e.g., 2-furfurylthiol from cysteine or 2-acetyl-1-pyrroline from proline, systematic studies correlating the overall odor of a given flavoring with the odor activity values (OAV: ratio of concentration to odor threshold) of the key odorants are scarcely performed. Extended knowledge on the structures and concentrations of key odorants formed, in correlation with the amounts of precursor present or the processing conditions, are other prerequisites enabling the manufacturer to optimize the flavor formation more towards the food flavor to be intensified or even to be mimicked. This publication summarizes own results aimed at characterizing potent aroma compounds in Maillard-type reactions using the odor-activity-value concept. Starting with data on the identification of the aroma compounds formed in binary mixtures containing one amino compound (cysteine, glutathione, proline) and one carbohydrate (ribose, rhamnose), the influence of the processing conditions on the yields of certain key odorants is discussed, based on knowledge about precursors and reaction pathways. Results obtained by thermal treatment of more complex precursor mixtures, such as yeast extracts, are presented with special emphasis on elucidating correlations between the concentrations of precursor amino acids present with the yields of key odorants formed.

1. INTRODUCTION

About 400000 years ago men have invented how to make fire and that was undoubtedly the beginning of food processing. It might be speculated that, after having recognized the benefits of heat in food preservation, our progenitors have, for sure, made use of the thermal treatment to generate the unique flavors of, e.g., roasted meat or cereal products.

Modern food production often requires the manufacturing of flavors by heating mixtures of certain pure ingredients occurring in foods, separately from the food itself. Therefore, there is a big interest in extending the scientific knowledge on flavor precursors and reaction pathways leading to flavor formation by thermal processes.

One of the most important flavor generating reactions in processed foods is the „Maillard-reaction" named after the French chemist Louis Camille Maillard. When heating a reducing carbohydrate and an amino acid in a model system, he observed the rapid development of brown color and odor. Studies correlating the overall odors of model Maillard systems with the amino acid used and, also, the processing conditions pointed to an important influence of the amino acid structure on the overall odor quality formed (1). Based on such findings, today this reaction is frequently applied in the manufacturing of the so-called reaction or processed flavors. Such flavorings are used to render certain aromas to a number of convenience foods, such as soups or snacks.

Although numerous studies have been performed aimed at identifying the volatiles generated by Maillard-type reactions, there is a lack of systematic studies (i) evaluating the flavor contribution of single aroma compounds to the overall aroma, (ii) correlating differences in the overall odor profiles with exact quantitative data on the key odorants generated and (iii) correlating the amounts of precursors present with the yields of the respective odorants formed after the thermal treatment.

Within the last 15 years a four step procedure based on the odor-activity-value concept has been developed by our group to identify those odorants mainly responsible for the overall aroma of a product (cf. review in 2).

A selection of ten very potent odorants, which have been confirmed as key odorants in several processed foods using the odor-activity-value concept is shown in Fig. 1. All of them are well-known Maillard-type reaction products and some are known to be formed from certain amino acids as precursors. E.g., the 2-acetyl-1-pyrroline is formed from proline and ornithine (3) whereas 2-acetyl-2-thiazoline is formed from cysteine in the presence of carbohydrates (4, 5).

One of the biggest challenges in reaction flavor development is to mimick a certain food aroma. However, because the development of these products is often done by empirical rather than by scientific approaches, it is often unclear whether changes in the overall odors are caused by the formation of new odorants or by quantitative changes in odorant composition. Furthermore, it often remains to be clarified which precursors or processing conditions favor the formation of certain potent odorants. Such data would, however, enable the manufacturers to generate the desired flavors based on knowledge about the concentrations present in the food itself compared with those present in the reaction flavor.

In the following, results of model studies performed in our group are reported elucidating possibilities to systematically modify the overall odors of reaction flavors based on chemical knowledge about precursors and their degradation pathways.

2. EXPERIMENTAL SECTION

2.1 Cysteine/carbohydrate models

Cysteine (3.3 mmol), dissolved in phosphate buffer (100 mL; pH 5.0; 0.5 mol/L) was reacted for 20 min in the presence of either rhamnose or ribose (10 mmol) at 145°C in an autoclave. Details on the isolation of the volatile fraction, GC/Olfactometry and the sensory evaluation are reported elsewhere (*4, 5*).

2.2 Yeast extracts

Either a commercial yeast extract (CYE; 20 g) or the freeze dried material isolated from a self-prepared yeast extract (SPYE; 20 g) was dissolved in water (60 mL) and thermally treated for 20 min at 145°C in a laboratory autoclave. After cooling, isotopically labeled internal standards were added and selected key odorants were quantified (*6*). Glucose or 2-oxopropanal, respectively, were added in a 1:1 or 10:1 molar ratio, respectively, based on the N-content of the yeast extract used.

3. RESULTS AND DISCUSSION

3.1 Cysteine as flavor precursor

3.1.1 Influence of the carbohydrate moiety on key odorants

In Fig. 2, spider web diagrams obtained by an overall sensory evaluation of two cysteine model systems reacted either in the presence of ribose or rhamnose are compared. The results clearly indicated a strong influence of the carbohydrate moiety on the overall odor generated. While intense roasty, meat-like note predominated in the cysteine/ribose mixture (Fig. 2; upper diagram), caramel-like and seasoning-like notes were most intense in the presence of rhamnose (Fig. 2; lower diagram). Application of an Aroma Extract Dilution Analysis (AEDA) in combination with the identification experiments (Table I) suggested 2-furfurylthiol, 3-mercapto-2-pentanone, 5-acetyl-2,5-dihydro-1,4-thiazine, 3-mercaptobutanone and 2-methyl-3-furanthiol as key contributors to the overall odor of the cysteine/ ribose mixture, whereas, due to their high FD-factors, the caramel-like smelling 4-hydroxy-2,5-dimethyl-3(2H)-furanone and 3-hydroxy-6-methylpyran-2-one with a seasoning-like odor contributed most to the odor of the cysteine/ rhamnose mixture (*4, 5*).

When applying AEDA, odor-active compounds are ranked according to their odor thresholds in air. Furthermore, the complete amount of a given odorant analyzed by HRGC is vaporized in the sniffing port. However, in a matrix, the odor activity of an odorant depends on its odor threshold in this matrix which is in fact displayed by its volatility (cf. review in *2*). A comparison of the odor thresholds of selected odorants in air and water (Table II) elucidates that the threshold ratios of water to air differ significantly. The data imply that, e.g., the odor contribution of 4-hydroxy-2,5-dimethyl-3(2H)-furanone relative to that of 2-methyl-3-furanthiol may be overestimated when applying AEDA. Furthermore, when considering the aroma contribution of single odorants, it has to be taken into account that losses during

136

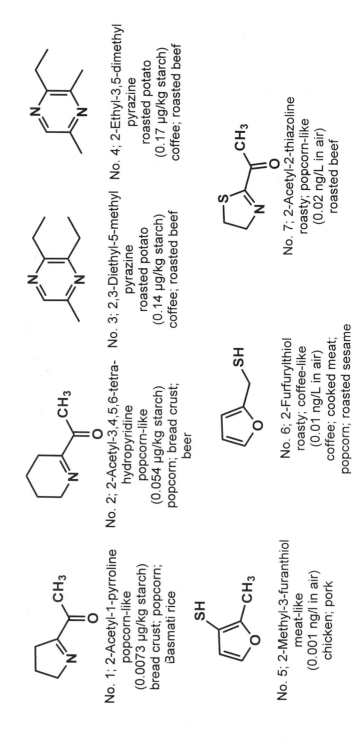

No. 1; 2-Acetyl-1-pyrroline
popcorn-like
(0.0073 µg/kg starch)
bread crust; popcorn;
Basmati rice

No. 2; 2-Acetyl-3,4,5,6-tetra-
hydropyridine
popcorn-like
(0.054 µg/kg starch)
popcorn; bread crust;
beer

No. 3; 2,3-Diethyl-5-methyl
pyrazine
roasted potato
(0.14 µg/kg starch)
coffee; roasted beef

No. 4; 2-Ethyl-3,5-dimethyl
pyrazine
roasted potato
(0.17 µg/kg starch)
coffee; roasted beef

No. 5; 2-Methyl-3-furanthiol
meat-like
(0.001 ng/l in air)
chicken; pork

No. 6; 2-Furfurylthiol
roasty; coffee-like
(0.01 ng/L in air)
coffee; cooked meat;
popcorn; roasted sesame

No. 7; 2-Acetyl-2-thiazoline
roasty; popcorn-like
(0.02 ng/L in air)
roasted beef

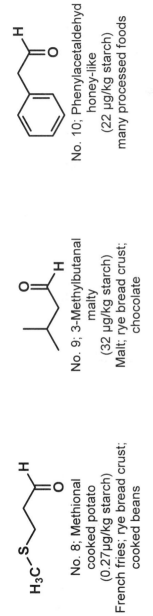

No. 8; Methional
cooked potato
(0.27 µg/kg starch)
French fries; rye bread crust;
cooked beans

No. 9; 3-Methylbutanal
malty
(32 µg/kg starch)
Malt; rye bread crust;
chocolate

No. 10; Phenylacetaldehyd
honey-like
(22 µg/kg starch)
many processed foods

*Figure 1. Intense odorants characterized as key aroma compounds in several foods
(odor thresholds in parentheses)*

138

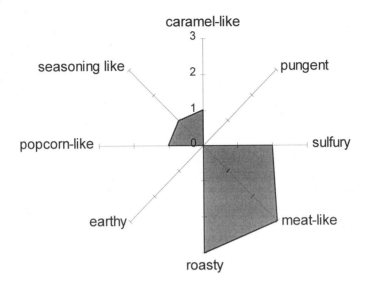

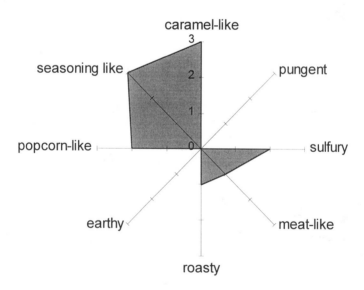

Figure 2. Spider web diagrams summarizing results of the odor evaluation of a thermally treated ribose/cysteine (upper) and a thermally treated rhamnose/cysteine mixture

Table I. Most intense odorants generated by thermal treatment of cysteine in the presence of ribose (A) or rhamnose (B). Source: Data from reference 4 and 5

Odorant	Odor quality	Flavor Dilution (FD) factor in	
		A	B
2-Furfurylthiol	roasty, coffee-like	1024	512
3-Mercapto-2-pentanone	catty, sulfury	512	128
2-Methyl-3-furanthiol	meat-like	256	<4
3-Mercapto-2-butanone	sulfury	128	32
4-Hydroxy-2,5-dimethyl-3(2H)-furanone	caramel-like	32	65536
3-Hydroxy-6-methylpyran-2-one	seasoning-like	<4	16384
5-Methyl-2-furfurylthiol	roasty, coffee-like	<4	2048
5-Acetyl-2,3-dihydro-1,4-thiazine	roasty, popcorn	256	512

Table II. Comparison of odor threshold of some key flavor compounds in air (ng/L) and water (µg/L)

Odorant	Odour threshold		Ratio
	air (ng/L)	water (µg/L)	Water/Air x 10^3
2-Methyl-3-furanthiol	0.005	0.007	1.4
2-Furfurylthiol	0.005	0.01	1.9
5-Acetyl-2,5-dihydro-1,4-thiazine	0.06	1.25	20.8
4-Hydroxy-2,5-dimethyl-3(2H)-furanone	1.0	60	60

isolation of the volatiles caused by, e.g., their different chemical stabilities may occur.

To overcome these drawbacks and to link AEDA results with the situation in a real food matrix, the concentrations of 15 key odorants were determined by stable isotope dilution assays in both models and their odor activity values (OAV: ratio of concentration to odor threshold) were calculated on the basis of odor thresholds in water. The results (*Hofmann and Schieberle, J. Agric. Food Chem., in preparation*), presented in part in Table III, confirmed that, in aqueous solution, 2-furfurylthiol and 2-methyl-3-furanthiol are by far the most important odorants in the cysteine/ ribose system. A similar importance was found for 4-hydroxy-2,5-dimethyl-3(2H)-furanone (caramel-like) and 3-hydroxy-6-methylpyran-2-one (seasoning-like) in the cysteine/rhamnose flavoring. However, in both systems, 3-mercapto-2-butanone and 5-acetyl-2,3-dihydro-1,4-thiazine showed lower OAV's than it might have been expected from the AEDA results (cf. Table I). This result, which is undoubtedly caused by differences in the odor thresholds of the compounds in either air or water, clearly indicates the need to confirm FD-factors by quantitative measurements when odor contributions of single aroma compounds are evaluated.

Table III. Concentrations, odor thresholds and odor activity values of key
odorants in a thermally treated (145°C; 20 min) aqueous cysteine/ribose (A) or
cysteine/rhamnose mixture (B)

Odorant	Odor thresholds (μg/L water)	Concentration (μg/L) in[a]		OAV[b]	
		A	B	A	B
2-Furfurylthiol	0.01	121	8	12100	800
3-Mercapto-2-pentanone	0.7	599	73	856	104
2-Methyl-3-furanthiol	0.007	198	8	28286	1143
3-Mercapto-2-butanone	3	342	141	114	47
4-Hydroxy-2,5-dimethyl-3(2H)-furanone	10	185	198000	19	19800
3-Hydroxy-6-methylpyran-2-one	15	<1	245300	<1	16353
5-Methyl-2-furfurylthiol	0.05	<0.1	65	<1	1354
5-Acetyl-2,3-dihydro-1,4-thiazine	1.3	424	401	340	321

[a] Concentrations are based on 33 mmol of cysteine and 100 mmol of the carbohydrate.
[b] Odor activity values were calculated by dividing the concentrations by the odor threshold in water.

Recent experiments showed (7) that a model flavor mixture consisting of 12 reference odorants in the same concentrations as they occur in the reaction mixtures showed nearly identical odors as compared to the complete Maillard reaction mixture (7). The identity of the overall odors was checked in terms of odor qualities perceived as well as of intensities of the single odor qualities.

3.1.2 Influence of the processing parameters
The processing conditions are known to significantly influence the overall odors generated in Maillard-type systems (8). Based on quantitative experiments, we could recently show (9) that compared to the aqueous reaction system (cf. Table III), significantly higher amounts of, in particular, 2-furfurylthiol were generated from a dry-heated (180°C; 6 min) cysteine/ribose-mixture. On the other hand other odorants, such as the 5-acetyl-2,3-dihydro-1,4-thiazine, were drastically decreased. To study the influence of temperature and time on odorant formation in more detail, a cysteine/ribose mixture was reacted under aqueous conditions at 100°C and the time course of the formation of several key odorants was followed. The results (Hofmann and Schieberle, unpublished) revealed (Table IV) that, with the exception of the 5-acetyl-2,3-dihydro-1,4-thiazine, the concentrations of the odorants were increased continuously with time. The most significant increase was observed within the first 6 hours at 100°C. Under these conditions, based on an amount of 10 mmol of ribose and 3.3 mmol of cysteine, by a factor of 13 higher yields of e.g., 2-methyl-3-furanthiol were obtained than in the mixture reacted for 20 min at 145°C (cf. Tables III and IV). The data indicate that heating such precursor systems at lower temperatures but for longer times may be useful in generating higher amounts of important key odorants.

Table IV. Influence of the processing time on the formation of selected
odorants from ribose and cysteine. Data from reference 16[a]

Odorant	Amount (μg) formed after			
	30	60	360	720
		min at 100°C		
2-Methyl-3-furanthiol	4.5	13.8	156	179
2-Furfurylthiol	2.0	3.1	110	132
3-Mercapto-2-pentanone	2.1	10.5	79	85
4-Hydroxy-2,5-dimethyl-3(2H)-furanone	<0.1	1.5	17.1	25.6
5-Acetyl-2,3-dihydro-1,4-thiazine	0.3	5.2	0.2	<0.1

[a] Ribose (10 mmol) and cysteine (3.3 mmol) were reacted in phosphate buffer (100 mL; 0.5 mol/L; pH 5.0) at 100°C in a laboratory autoclave.

3.1.3 Influence of the addition of lipids

It has been reported that additions of soya lecithine to a cysteine/ribose mixture changed the overall odor to a more beef broth-like aroma and, also, significant changes in the composition of the volatile fraction were observed (*10*). When repeating these model studies under somewhat different reaction conditions, we also observed the same difference in the overall odors. Application of the Aroma Extract Dilution Analysis on two extracts prepared from a thermally treated cysteine/ribose solution and a ternary mixture containing cysteine, ribose and soya lecithine indicated (*Hofmann and Schieberle, unpublished*) that addition of soya lecithine decreased the FD-factors of all three thiols with 2-methyl-3-furanthiol being decreased by a factor of more than 30 (Table V). On the other hand, (E,Z)- and (E,E)-2,4-decadienal, showing deep-fat fried odor qualities, appeared as additional, most odor-active compounds in the mixture containing lecithine. Quantitative measurements and recombination experiments would, however, be necessary to confirm that the reduction of the thiols and the formation of the dienals are the main reason for the flavor differences caused by the addition of soya lecithine.

3.2 Glutathione as flavor precursor

As shown in Fig. 3, several reactive intermediates may be generated by a so-called *Strecker* degradation of cysteine, such as cysteamine, hydrogensulfide, ammonia or mercaptoacetaldehyde. Besides the free amino acid, the tripeptide glutathione is also used as a sulfur source in the production of reaction flavors. However, based on chemical mechanisms, from the „bound" cysteine in glutathione only hydrogensulfide, but not cysteamine can be liberated. This difference in the possible degradation pathways of free cysteine and „bound" cysteine in glutathione should, consequently, influence the yields of certain sulfur containing odorants.

Recently, we could show that 2-furfurylthiol is generated in relatively high yields from the reaction of furan-2-aldehyde and hydrogensulfide, and we have proposed the formation pathway shown in Fig. 5. 5-Acetyl-2,3-dihydro-1,4-thiazine is likely to be formed from the reaction of 2,3-butanedione and cysteamine (*11*).

Table V. Comparative Aroma Extract Dilution Analysis of an aqueous cysteine/ribose mixture thermally treated in the presence or absence of soya lecithine[a]. Source: Data from reference 16

Odorant[b]	FD-factor	
	without	with lecithine
2-Furfurylthiol	4096	1024
3-Mercapto-2-pentanone	2048	256
2-Methyl-3-furanthiol	1024	32
5-Acetyl-2,3-dihydro-1,4-thiazine	1024	512
3-Mercapto-2-butanone	512	32
Bis-(2-methyl-3-furyl)-disulfide	512	16
(E,Z)-2,4-Decadienal	<1	128
(E,E)-2,4-Decadienal	<1	512

[a] Ribose (10 mmol), cysteine (3.3 mmol) and soya bean lecithine (0.5 g) were reacted in phosphate buffer (100 mL; pH 5.0; 0.5 mol/L) for 20 min at 145°C.
[b] Odorants showing FD-factors ≥128 are given.

Table VI. Comparison of the concentrations of selected key odorants in a ribose/cysteine and a ribose/glutathione mixture[a]

Odorant	Conc. (µg) per 3.3 mmol of	
	cysteine	glutathione
2-Methyl-3-furanthiol	19.8	18.5
2-Furfurylthiol	12.1	12.9
3-Mercapto-2-pentanone	59.9	25.1
5-Acetyl-2,3-dihydro-1,4-thiazine	42.5	<0.1
2-Acetyl-2-thiazoline	7	<0.1

[a] Ribose (10 mmol) and cysteine (3.3 mmol) or glutathione (3.3 mmol), respectively, were reacted in phosphate buffer (100 mL; 0.5 mol/L; pH 5.0) for 20 min at 145°C. Source: Data from reference 16.

Consequently, based on the proposed reaction mechanism (11), 5-acetyl-2,3-dihydro-1,4-thiazine and 2-acetyl-2-thiazoline should not be formed when cysteine is replaced by glutathione in the model system. To prove this assumption, model mixtures of cysteine/ribose and glutathione/ribose were reacted under aqueous conditions (for details see Table VI) and the five sulfur containing compounds listed in the table were quantified by stable isotope dilution analyses (Hofmann and Schieberle, unpublished results). The data indicate that the concentrations of 2-furfurylthiol and 2-methyl-3-furanthiol were not significantly affected when cysteine was substituted by glutathione in the reaction with ribose. This is well in line with the reaction pathway shown in Fig. 4, because H_2S, as the possible precursor of both odorants, can be formed from both, cysteine and glutathione. However, the two S-containing heterocycles 5-acetyldihydrothiazine and 2-acetyl-2-thiazoline were not formed from glutathione corroborating cysteamine, which cannot be formed from glutathione, as precursor of the latter two odorants.

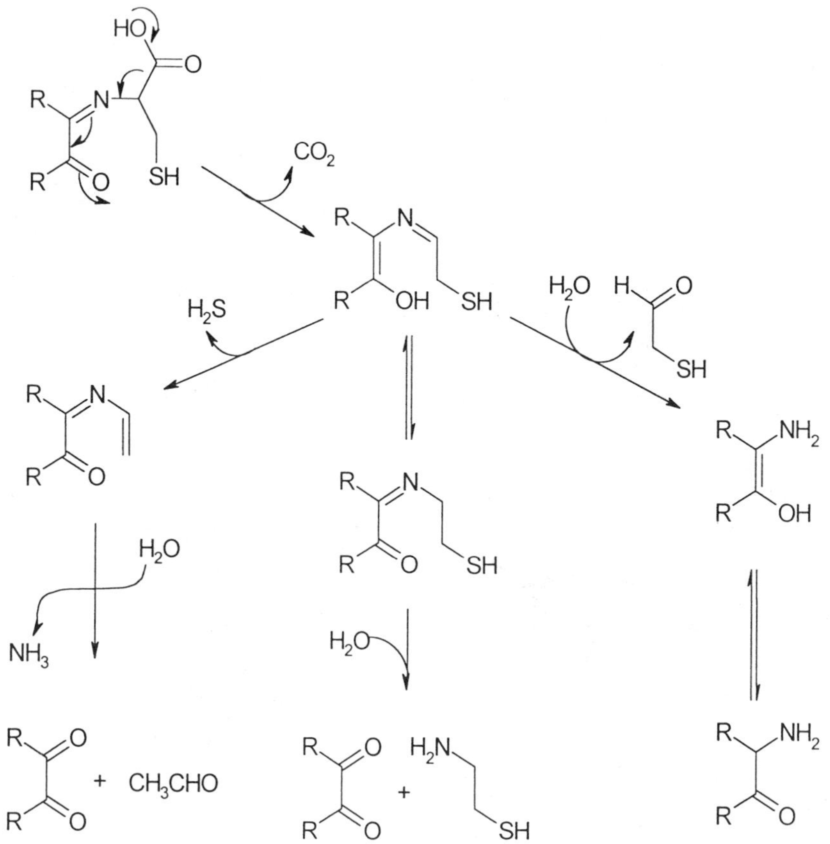

Figure 3. Possible pathways of the Strecker degradation of cysteine initiated by α-dicarbonyls

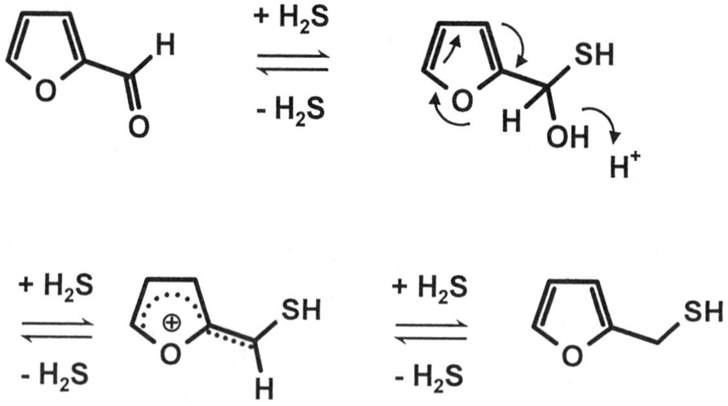

144

Figure 4. Formation of 2-furfurylthiol via reductive sulfhydrylation of furfural by H₂S

Table VII. Key odorants (FD $\geq$256) generated in a thermally treated proline/glucose reaction mixture[a]

Odorant	Odor quality	FD-Factor
2-Acetyl-1-pyrroline	popcorn-like	16384
2-Propionyl-1-pyrroline	popcorn-like	512
2-Acetyl-3,4,5,6-tetrahydropyridine	popcorn-like	2048
4-Hydroxy-2,5-dimethyl-3(2H)-furanone	caramel-like	8192
2-Acetyl-1,4,5,6-tetrahydropyridine	popcorn-like	2048

[a] The reactants were mixed with silica and thermally treated for 10 min at 160°C in a metal bloc. Source: Data from reference 12.

4. PROLINE AS PRECURSOR

In a further experiment, the amino acid proline was dry-heated for 10 min at 160°C in the presence of glucose, yielding a very intense popcorn, bread-crust like odor. Application of the AEDA on the volatile fraction isolated by extraction with diethyl ether and sublimation in vacuo (Table VII) revealed 2-acetyl-1-pyrroline (popcorn-like), followed by 4-hydroxy-2,5-dimethyl-3(2H)-furanone (caramel-like) and the two tautomers of 2-acetyltetrahydropyridine (popcorn-like) as the most odor-active compounds among the odorants identified (*12*). The results confirmed previous results by Roberts and Acree (*13*), establishing proline as an important precursor amino acid in the generation of popcorn- or cracker-like aromas and elucidated 2-acetyl-1-pyrroline and 2-acetyltetrahydropyridine as the key odorants in the overall aroma. Systematic studies on their formation revealed that both odorants are formed from the same degradation product of proline, 1-pyrroline (*14*). Further data showed that it is the carbohydrate degradation product, which determines the reaction pathway. Reaction of 1-pyrroline with hydroxy-2-propanone exclusively yields 2-acetyltetrahydropyridine, whilst 2-oxopropanal gives rise to 2-acetyl-1-pyrroline (Fig. 5).

5. YEAST EXTRACT AS FLAVOR PRECURSOR MIXTURE

5.1 Characterization of key odorants

In the industrial manufacturing of reaction flavors very often complex mixtures of amino acids are used, which are obtained by either hydrolysis of bakers or brewers yeast or hydrolysis of plant proteins, e.g. wheat gluten. Upon thermal treatment, yeast extracts are known to generate predominantly meat-like odor notes, however, it is yet unclear which compounds mainly contribute to the typical aroma of such processed yeast extracts. By application of the AEDA on a commercial yeast extract (CYE) we could recently show (*6*) that 2-furfurylthiol followed by 4-hydroxy-2,5-dimethyl-3(2H)-furanone and 2- and 3-methylbutanoic acid showed the highest FD-factors among the odor-active compounds identified after thermal treatment in aqueous solution (20 min; 145°C).

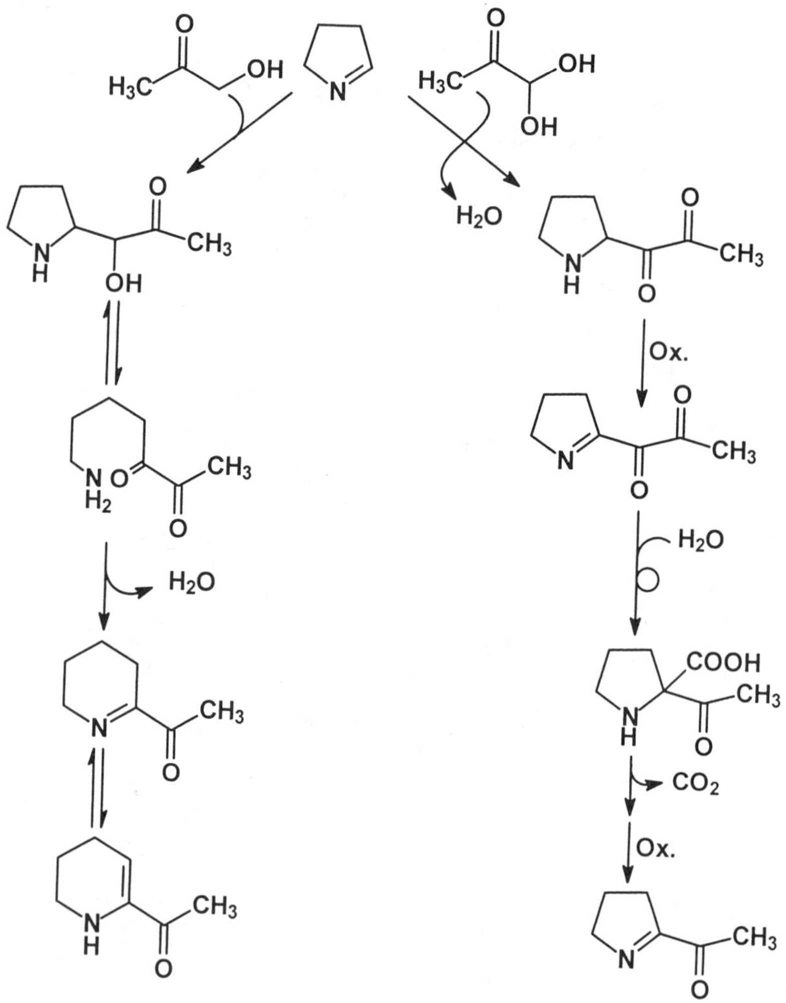

Figure 5. Formation pathways yielding 2-acetyl-1-pyrroline and 2-acetyltetrahydropyridine from their common precursor 1-pyrroline

Table VIII. Concentrations and odor activity values of selected key odorants in a thermally treated (20 min; 145°C) commercial yeast extract (CYE) and a self-prepared yeast extract (SPYE)

Odorant (odor threshold)	Conc. (µg/kg)		OAV	
	CYE	SPYE	CYE	SPYE
2-Methyl-3-furanthiol (0.007)	530	22	75714	3140
2-Furfurylthiol (0.01)	580	29	58000	2900
3-Methylbutanal (0.4)	1456	617	3640	1543
Methional (1.8)	1095	287	608	159
Phenylacetaldehyde (4.0)	1250	n.d.	313	—
.				.
.				.
.				.
2-Acetyl-1-pyrroline (0.1)	2.2	6	22	60

For comparison, a baker's yeast extract was then prepared by autolysis of bakers yeast under laboratory conditions (SPYE; 6). The thermal treatment resulted in the formation of the same odorants as identified in the commercial extract, however, different FD-factors were determined. For example, the FD-factor of 2-furfurylthiol was significantly lower in this extract.

A quantitation of 18 of the most odor-active compounds in both extracts followed by a calculation of their odor activity values (15) revealed six compounds with OAV's >300 in the CYE (Table VIII). Among them 2-methyl-3-furanthiol and 2-furfurylthiol were shown to be the most important aroma contributors. It is interesting to note that 2-methyl-3-furanthiol could not be detected by the AEDA experiments, but showed a high OAV in the quantitative experiments. This is undoubtedly caused by the instability of this odorant during the work-up procedures, which was overcome by application of the stable isotope dilution analysis.

5.2 Quantitative correlations of odorants and amino acid precursors

As already discussed above, cysteine is an important precursors of the two thiols. Because, compared to the SPYE, much higher concentrations of both odorants were measured in the CYE (Table VIII) it might be assumed that the concentration of cysteine was higher in the CYE. This assumption was confirmed by a quantitation of cysteine in both extracts indicating a much lower concentration in the SPYE (Table IX).

As examplified for phenylalanine in Fig. 6, the oxidative decarboxylation of amino acids initiated by an α-dicarbonyl compound, the so-called *Strecker* degradation, generates flavor-active aldehydes, such as phenylacetaldehyde. It is interesting to note that, compared to the relatively high amounts of the corresponding precursor amino acids leucine (for 3-methylbutanal), phenylalanine and methionine (for methional) present in the yeast extracts (cf. Table IX), the yields of the three *Strecker* aldehydes were comparatively low (Table VIII). For example, from 14 g of leucine/kg yeast extract (Table IX) only 1.46 mg of 3-methylbutanal (Table VIII) were generated (equals to about 0.17 %).

Table IX. Concentrations of selected free amino acids in the commercial yeast
extract (CYE) and the self-prepared yeast extract (SPYE)

Amino acid	Conc. (mg/kg dry weight)	
	CYE	SPYE
Cysteine	21300	210
Leucine	14003	10280
Methionine	4280	2534
Phenylalanine	9364	5447
Proline	3750	1550
Ornithine	2700	1580

Table X. Concentrations of odorants formed by a thermal treatment of a
commercial yeast extract (CYE) - Influence of additions of glucose (B) or 2-
oxopropanal (C)

Odorant	Conc. (μg/kg)		
	A^a	B	C
3-Methylbutanal	1456	3418	9391
Methional	1095	1707	29068
Phenylacetaldehyde	1250	3056	18510
2-Methyl-3-furanthiol	530	122	158
2-Furfurylthiol	580	13	122

[a] Control without carbohydrate addition.

To study the influence of additions of carbohydrates on the yields of key
odorants, yeast extracts were dissolved in water containing either an equimolar
amount glucose or one tenth of 2-oxopropanal. In the thermally treated mixtures the
concentrations of the *Strecker*-aldehydes were then quantified by means of stable
isotope dilution analyses. The results indicated (Table X) that glucose only slightly
enhanced the concentrations of the three aldehydes (cf. B with A; Table X). On the
other hand, the amounts of the two thiols were significantly reduced (C; Table X).
The same was true when glucose was substituted by 2-oxopropanal in the reaction
with the yeast extract. However, the yields of the *Strecker* aldehydes formed in the
presence of 2-oxopropanal were much higher. E.g., methional was increased by a
factor of nearly twenty-seven compared to a factor of three in the presence of
glucose (cf. C with A; Table X). In this experiment, based on the amount of the
precursor methionine present (cf. Table IX), the yields of methional were calculated
to be about one percent.

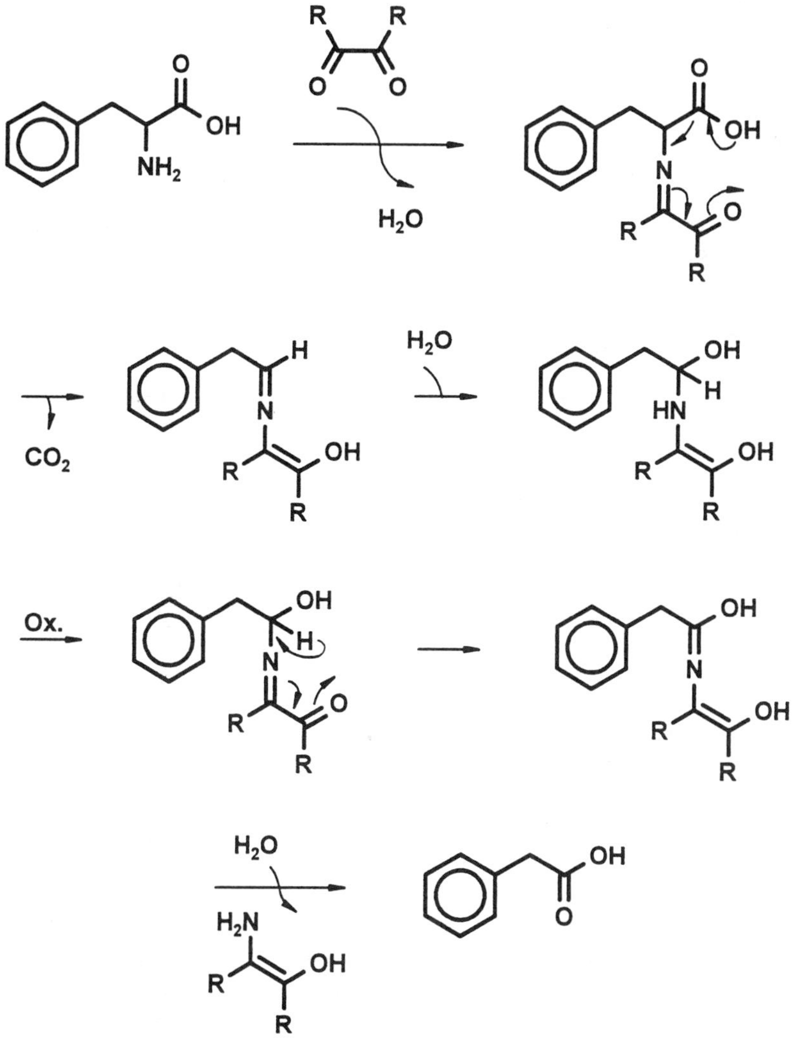

Figure 6. Strecker degradation of phenylalanine leading to phenylacetaldehyde

6. CONCLUSIONS

The results propose that only a few, very potent odorants are formed when amino acids and carbohydrates are reacted at higher temperatures. It is, therefore, understandable that these compounds, due to their high odor potencies, are often identified as important odorants also in processed foods or reaction flavors. For example, the coffee and stewed beef odorants 2-furfurylthiol and 2-methyl-3-furanthiol were confirmed as main odor contributors in the cysteine containing mixtures, whereas from proline, the key bread crust and popcorn odorants 2-acetyl-1-pyrroline and 2-acetyltetrahydropyridine were generated. Thermal treatment of more complex precursor mixtures, such as yeast extracts, however, revealed that the composition and yields of key odorants are often changed compared to binary mixtures.

The development of more sophisticated ways in reaction flavor manufacturing, based on extended knowledge on, e.g., the yields of key odorants, reaction pathways and reaction intermediates should, consequently, be an interesting approach to generate the desired overall aroma of a given food by means of Maillard-type reactions.

7. LITERATURE CITED

1. Herz, K.O.; Chang, S.S. *Adv. Food Research*, **1970**, *18*, 1.
2. Schieberle, P. In Characterization of Food: Emerging Methods, Goankar, A.G., Ed.; Elsevier, **1995**, pp. 403.
3. Schieberle, P. *Z. Lebensm. Unters. Forsch.*, **1990**, *191*, 206.
4. Hofmann, T.; Schieberle, P. *J. Agric. Food Chem.* **1995**, *43*, 2187.
5. Hofmann, T.; Schieberle, P. *J. Agric. Food Chem.* **1997**, *45*, 898.
6. Münch, P.; Hofmann, T.; Schieberle, P. *J. Agric. Food Chem.* **1997**, *45*, 1338.
7. Schieberle, P.; Hofmann, T. In Flavour Science, Recent Developments; Taylor, A.J.; Mottram, D:S. (eds.) The Royal Society of Chemistry, **1996**, pp. 175.
8. Lane, M.J.; Nursten, H. In The Maillard Reaction in Food and Nutrition, ACS Symposium Series 215; American Chemical Society: Washington, DC, **1983**, pp. 141.
9. Schieberle, P.; Hofmann, T. In: Flavor Analysis - Developments in Isolation and Characterization, ACS Symposium Series 705, American Chemical Society, Washington, DC, **1998**, pp. 320-330.
10. Whitfield, F.; Mottram, D.S.; Brock, S.; Puckey, D.J.; Salter, L.J. *J. Sci. Food Agric.* **1988**, *42*, 261.
11. Hofmann, T.; Haeßner, R; Schieberle, P, *J. Agric. Food Chem.* **1995**, *43*, 2195.
12. Hofmann, T.; Schieberle, P. *J. Agric. Food Chem.* **1998**, *46*, 2721.
13. Roberts, D.D.; Acree, T.E. In Thermally Generated Flavors, Parliment, T.H.; Morello, M.J.; McGorrin, R. (eds.) American Chemical Society: Washington, D.C., **1994**, pp. 71.
14. Hofmann, T.; Schieberle, P. *J. Agric. Food Chem.*, **1998**, *46*, 2270.
15. Münch, P.; Schieberle, P. *J. Agric. Food Chem.* **1998**, *46*, 4695.
16. Hofmann, T.; Schieberle, P. to be submitted for publication.

Chapter 11

Flavor Release and Flavor Perception

A. J. Taylor, R. S. T. Linforth, I. Baek, M. Brauss,
J. Davidson, and D. A. Gray

Samworth Flavor Laboratory, Division of Food Sciences,
University of Nottingham, Sutton Bonington Campus,
Loughborough LE12 5RD, United Kingdom

The relationship between the amount of a single flavor chemical (the stimulus) and the flavor perception (the response) that it invokes when food is eaten, has been studied for many years. The so-called Psychophysical Laws provide a simple mathematical way of expressing the relationship between the amount of a flavor compound and the flavor intensity it generates. These Laws have been developed using carefully defined conditions of presentation and most values have been obtained with non-volatile taste compounds. A modification to the Power Law, which took into account the effect of adaptation, was proposed by Overbosch (1) but the concept has not been extensively tested. Since the stimulus-response relationship for single volatiles is not entirely clear, attempts to understand the relationship between the mixtures of components in a typical commercial flavoring and the quality and intensity of the overall flavor perceived have been fraught with difficulty. A rigorous scientific basis for understanding the relationship would be useful in a variety of applications, such as the predictive modelling of perceived flavor from the flavor composition of a food or a better understanding of the olfactory reception process.

The proposition in our research group has been that appropriate analyses underpin the whole process. Novel methods have been developed to measure the chemical stimuli that produce aroma and taste perception, close to the site at which stimulation occurs. Using model foods and single flavor compounds, we have investigated the role of the "temporal dimension" on sensory perception as well as the interactions that occur between non-volatile and volatile compounds to modify overall flavor perception. Lastly, the relative contribution of flavor release during eating, on overall flavor perception, has been studied by sampling the stimuli in nose over the whole "eating experience".

The study of the link between the chemical stimuli that make up flavor and the sensory responses they evoke when food is consumed, has been pursued for many years. However, the main thrust of flavor research, following the introduction of Gas Chromatography (GC) was the determination of the volatile composition of foods. Stewart (2) in an editorial in Food Technology, encouraged researchers to "establish *which* of the compounds isolated are responsible for *what* sensory properties" and in 1994, Williams (3) presented a paper entitled "Understanding the relationship between sensory responses and chemical stimuli. What are we trying to do? The data, approaches and problems" which suggested that there was still much to do some 30 years after Stewart's plea. The classic Laws of Psychophysics date from the mid 1800s (see review 4) but a wider view of olfactory and taste psychophysics can be found in Lawless (5) and Halpern (6), respectively. All studies on the subject basically consist of four key steps.

- Develop appropriate analyses to measure the stimuli
- Measure sensory responses caused by the stimuli e.g. (time intensity or overall response)
- Devise mathematical (or other) relationships between stimuli and response
- Test models to validate the relationship

The belief in our research group is, that appropriate analyses underpin the whole process and, if the analysis delivers poor quality data, then the success of the other steps is compromised. The definition of appropriate analysis is that stimuli should be sampled close to the receptor at which they act. Since the sensory response of aroma is often the overriding component of flavor, much work has focused on aroma. To relate perceived aroma to volatile concentration, the headspace profile over an intact food sample has often been measured. However, for many foods, the physical changes that occur during eating cause major changes to the volatile profile found in mouth compared to the headspace profile. When volatiles are transported to the nose, further dilution and adsorption effects occur, which again alters the volatile profile. These aspects have been reviewed previously (7). A consequence of this type of sampling is that the concentrations of the volatiles are low and sensitive analyses are required. The various techniques available for trace volatile analysis have been discussed by Taylor & Linforth (8).

Now that suitable analyses are available, we have applied them to study other aspects of the stimuli-response relationship. For instance, do volatiles reach the olfactory receptors by the ortho- or retro- nasal route? Understanding this aspect would determine whether headspace or nosespace sampling was appropriate. Secondly, sensory Time Intensity (TI) experiments show clearly that sensory response changes with time but are there corresponding changes in the volatile signal with time? The role they play in perception is not known. Thirdly, it is known that flavor compounds tend to interact to cause perceptual changes (e.g. salt enhances flavor) but is it a physical effect (salting out), due to signal processing after the receptors, or a combination of both effects? Lastly, what is the relative importance of the various volatile flavor signals in determining overall flavor perception? With many foods, the first olfactory signal is sensed orthonasally, as the food is sniffed prior to consumption. A second signal is received during eating. Are these signals different and what is the relationship between them?

In this paper, we put forward some reasons why the relationship between stimulus and response is still not fully understood. They can be summarized as follows:-

- Stimuli should be sampled in the appropriate place
- Stimuli change with time
- There are interactions with other stimuli (especially from the taste sensors)
- Flavor stimuli in mouth or in nose during eating may only represent one part of the overall flavor experience.

Work in this area is at an early stage and the data presented give an illustration of the approach adopted in our laboratory. Extensive testing in a variety of foods and with a range of flavorings will be required before definitive answers can be produced.

Sampling flavor signals in the appropriate place

The perception of flavor by humans is normally sensed as a single entity (5). For instance we talk about strawberry flavor rather than assigning the flavor to the individual components e.g. a combination of acidity, sweetness, esters and furaneol. Describing perceived flavors in terms of the individual components is extremely difficult and Laing (9) claimed that even trained sensory panellists could not differentiate more than 3 or 4 aroma compounds when they were simultaneously presented. Experienced wine tasters provide one of the few examples of a situation where the overall flavor of a food can be expressed in terms of its flavor components, and in terms of the way the flavor changes with time.

The components of flavor are mainly chemical in nature (although physical stimuli like temperature also play a role), and they stimulate sensors in the mouth and nose. The stimulation produces a complex pattern of signals, which are processed locally and centrally to produce the sensation of flavor perception (10). There are obviously a number of separate, but sequential, processes occurring and it is important to consider what occurs at each stage and then devise appropriate measurements to define the signal.

Flavor perception can be crudely defined as:-

Flavor perception = aroma + taste + mouthfeel + texture + pain/irritation

Ideally, to characterise a flavor, it is necessary to measure **all** these parameters. As explained in the Introduction, aroma has been the subject of much research because it is often the dominant flavor component and methods for determining volatile compounds associated with flavor are well documented. Although it is convenient to focus on aroma, the contribution of the other components should not be forgotten as interactions between the components may be important in defining the flavor perceived. This is discussed later.

Methods are available for determining the **total volatile composition** of foods in a variety of foods by obtaining a volatile extract from the food, which is then analyzed by Gas Chromatography-Mass Spectrometry (GC-MS). Since aroma is sensed only when the volatile compounds enter the gas phase, much attention has been focused on the profile and concentration of volatile compounds in the **headspace** above a food. However, food undergoes major changes during eating (7,11,12) and this invariably affects both the profile of volatiles and, therefore, the amounts of volatiles that encounter the olfactory receptors. When considering the link between chemical stimuli and perception, it is appropriate to measure the concentration of volatiles found at the sensors rather than the concentration of volatiles in intact food or in the headspace above intact food.

Various methods have been developed to measure the concentration of volatiles close to the olfactory epithelium. Ideally, methods that measure the volatile concentration directly in real time are needed (8). There are difficulties in developing analyses with sufficient sensitivity to measure volatiles directly at their odor thresholds, which are in the ppb to ppt range. Trapping of expired air from people eating food, followed by GC-MS analysis of the samples, overcomes the sensitivity limitations and several research groups have reported success with this approach (13-16). However, to obtain a release curve with sufficient points entails many GC-MS runs and it can take several days (or weeks) of work to produce a single trace. A direct Mass Spectrometric method was devised by Soeting and Heidema (17), which gave breath by breath traces but sensitivity was not really sufficient, with detection in the high ppm (mg/kg) range.

Despite their limitations, these experiments clearly demonstrated that:

- *The volatile profile in expired air from the nose during eating is often very different from the headspace profile although it depends on the type of food under consideration*

- *The volatile profile in the nose actually changes over the short period of time that food is eaten and this raises the question of whether it is not just the type and amounts of compound present in the nose that define the aroma signal or whether there is also a temporal dimension to it*

The real time Atmospheric Pressure Ionization (API)-MS technique developed in our lab (18) is based on earlier work (19) and can measure volatile concentration on a breath-by-breath basis. It can detect volatile compounds at concentrations in the ppb to ppt (by volume) range providing sensitivity to measure about 80% of volatiles at their odor threshold (*20*). Since it resolves the incoming mixture of volatiles solely on the basis of mass (no chromatography is involved) it is incapable of separating compounds with the same mass or which produce ions of the same mass. Isomers, for instance cannot be differentiated. Since the stereoisomers of some compounds exhibit very different aromas, this limits the general applicability of the technique. However, the technique can still yield much valuable information and our approach has been used to study some fundamental aspects, using volatiles that can be resolved by API-MS.

The collection of expired air involves resting one nostril on a small plastic tube, through which expired air passes, and from which a portion of air is continuously sampled into the API-MS (Fig 1). On exhalation, expired air from the nose is sampled, on inhalation, laboratory air is sampled, giving rise to a series of peaks and troughs (Fig 2).

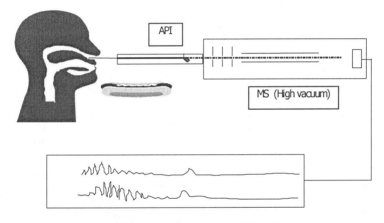

Breath by breath trace for up to 20 volatiles

Figure 1 Schematic of API-MS and breath collection

To ensure people are breathing through their nose, (and that their breathing pattern is not affected by the distractions of the laboratory environment) it has been our custom to monitor acetone on the breath. Acetone is present in blood, crosses through the lungs to the air and is then exhaled. It is therefore a marker for exhalation (Fig 2). If other volatile compounds are monitored breath by breath, those released in-mouth will be transported to the gas stream in-nose by the retronasal route and will therefore be in phase with the acetone trace. Any volatile that enters the nose by the orthonasal route, however, will be out of phase with the acetone pattern. Careful scrutiny of the breath by breath traces can provide an estimate of the contribution ortho- and retro- nasal routes are making to the volatiles present in-nose when foods are consumed. For some foods, particularly hot beverages (coffee, tea and chocolate), there is the possibility of volatiles being sampled by both routes. Thus API can indicate not only the concentration of volatiles in-nose on a breath by breath basis but give some clue as to their origin.

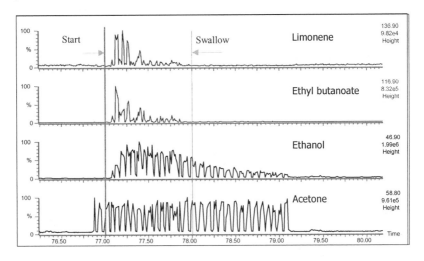

Figure 2. Release of selected volatiles from a fresh orange segment during eating

Measuring the concentration in-nose can also provide a useful check when carrying out other procedures like sensory sniff tests. Lawless (5) commented that the variation in aroma perception for tests using sniff bottles was higher than when an olfactometer was used to deliver volatiles to the nose. Experiments in our laboratory have shown that the headspace above a solution in a bottle contains an even concentration of volatiles at equilibrium but, when the headspace is diluted with air, there are substantial variations in concentration unless the gas phase is vigorously mixed. Coefficients of variation around 50 to 80% have been noted (unpublished data) and, given the nature of sniff bottles, similar variation might be expected with them. Since in-nose sampling is unobtrusive, it can be used simultaneously with sensory analysis and the actual concentration of volatile in-nose plotted against sensory response.

Applications of in-nose volatile monitoring

Several food systems have been studied to determine the effect of reformulation on volatile release in mouth. Since there is a demand for low fat, low sugar, high fibre foods, food manufacturers need to reformulate certain products, a task that invariably leads to flavor (and texture) changes, which are normally undesirable in nature. Using API-MS to monitor volatile release in-nose gives objective comparative data on the volatile release profile in the original and in the reformulated food. Figure 3 shows the anethole release from two model biscuit systems containing regular (18%) and low (4%) amounts of fat. Anethole was added to each dough at 250 mg/kg, prior to baking but the anethole amounts after baking were significantly different at 40mg/kg in the regular fat and 19 mg/kg in the low fat biscuits. During eating, the anethole concentration in-nose from the low fat biscuit was significantly higher than the regular fat sample (120 compared to 50 mg/kg) even though the concentration in the low fat biscuit was roughly half that of the regular fat sample.

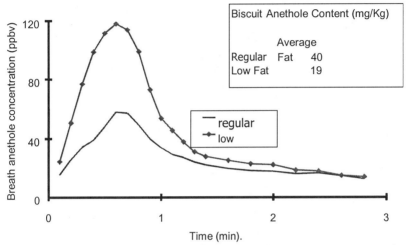

Figure 3. Release of anethole in mouth for low (4%) and regular (18%) fat biscuits

Further information on the comparative release of the added anethole and the thermally produced methyl butanals was obtained by using API-MS to measure headspace above a dry biscuit, which was then hydrated to levels found in mouth (Fig 4). The initial period showed that anethole was readily released into the headspace during the equilibration period while methyl butanals were not. As soon as water was added, methyl butanals were released. Further experiments demonstrated that the release of anethole was mainly driven by temperature while methyl butanals were driven by the degree of hydration. (Note: since API-MS cannot differentiate positional isomers, it was not possible to separate the 2- and 3- methyl butanal present in the biscuits. The generic term methyl butanals therefore has been used). Because API-MS can generate data in real time and simultaneously follow the release of up to 20 volatiles, experiments such as those described in Figs 3 and 4 are relatively quick to carry out. The data obtained can give information on the release of particular volatiles and suggest potential mechanisms for release. The release of volatiles from regular and low fat yogurts has also been investigated (21).

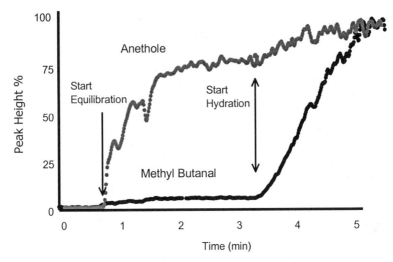

Figure 4. Release of added anethole and thermally generated methyl butanals from biscuits hydrated in a model system. Anethole was released into the headspace at the start of the experiment; methyl butanals were only released during hydration.

Temporal dimension to flavor perception

As stated previously, the Laws of psychophysics do not take into account the time that the stimulus is in contact with the receptors. However, sensory studies using Time Intensity analysis have shown that the aroma of foods changes with time and this suggests that the concentration of volatile at the receptors must also change. One of the key questions is whether this change in volatile concentration is important for the relationship between stimulus and response. One piece of relevant information is the fact that the olfactory system can become adapted to an aroma until, regardless of the magnitude of the stimulus, there is no response. People who have to work in odorous atmospheres adapt relatively quickly so that they are unable to smell the background odor. Another key question is whether adaptation is sufficiently fast to have a significant effect during the short time that food is in the mouth. Inspection of the literature available suggests that some compounds have short adaptation periods, while others are longer.

Overbosch (*1*) examined the situation that occurs when a constant level odor stimulus was applied to the olfactory receptors. He also formulated an expression to predict the response taking into account adaptation. It is basically the Power Law relationship but the stimulus is now the actual stimulus minus the adaptation that has occurred. The ideas proposed by Overbosch were developed in a series of papers (*22-*

24) and some experimental evidence was presented to support the hypotheses inherent in the adaptation model.

Experiments in our laboratory have been carried out using a simple gelatin-sucrose gel system that can be easily manipulated to deliver a single volatile (furfuryl acetate) at different rates. Trained panelists ate the gels and awarded them a score for flavor intensity while the concentration in-nose was monitored for the duration of eating (*25*). Fig 5 shows the relationship between sensory score and maximum in-nose concentration of furfuryl acetate during eating. Fig 6 shows a plot of rate of release against sensory score. The sensory scores vary in the range 3 to 9 and the scale was anchored using an unflavored sample to represent a score of 0 and a gel with 300 ppm furfuryl acetate representing a score of 10. The change in I_{max} values was in the range of 120 to 180 ppbv (nanoliter volatile per liter air). This seems a rather low range as many people consider that the volatile concentration needs to be doubled to provide a detectable sensory difference.

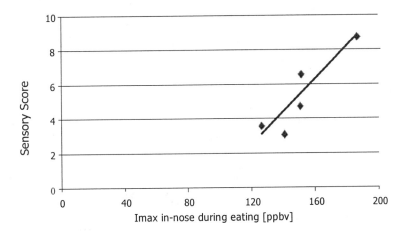

Figure 5. Plot of maximum in-nose concentration of volatile (Imax) against sensory score for gelatin/sucrose gels containing a single volatile (furfuryl acetate). Each point is the mean value from 16 panelists.

When the rate of release was plotted against sensory score (Fig 6), the change in rate of release spanned the range 120 to 550 ppbv/min. Interpretation of these results is difficult since both rate of release and I_{max} are related but there are some attractions in suggesting that the rate of release is a better indicator of sensory score than *Imax*. Further analysis of the data is required to establish if rate of release and adaptation do actually affect the sensory response.

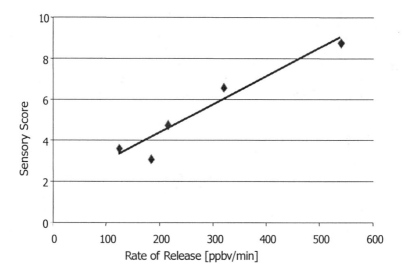

Figure 6. Plot of rate of release of furfuryl acetate from gelatin/sucrose gels against sensory score. Each point is the mean value from 16 panelists.

Overbosch (*1*) predicted that adaptation should have no effect on the sensory Time to Maximum Intensity (*Tmax*) parameter although his model was based on a constant stimulus rather than a changing stimulus. Experiments have been carried out in our laboratory with different volatiles in a gel system that provides different delivery rates. Panelists were not constrained in their chewing patterns and the result was a wide range of *Tmax* values. Fig 7 shows a plot where the ratio of *Tmax* for the sensory and volatile release parameters (sensory *Tmax* divided by volatile *Tmax*; the *Tmax* ratio) is plotted against the volatile *Tmax*. If Overbosch's prediction is correct, one might expect a constant value for the *Tmax* ratio, irrespective of when the volatile *Tmax* occurs. Despite the scatter, there was a clear trend with *Tmax* ratios above 1.0 at short volatile *Tmax* times and ratios below 1.0 at longer volatile *Tmax* values. This means that at short eating times, the volatile concentration is reaching its maximum value before sensory perception reaches its maximum value. One can hypothesize that the volatile concentration changes so rapidly in these situations that the sensory perception lags behind. Another way of expressing this lag idea is that the process of olfactory recognition takes a finite time and, over the short periods involved (up to 0.6 min) the nose is still double checking that it really is detecting an aroma, which causes a significant lag in sensory perception. Conversely, at volatile *Tmax* values greater than 0.6 min, sensory perception keeps pace with the increase in volatile concentration initially but adaptation now becomes significant, with the result that the sensory intensity decreases due to adaptation before the maximum volatile concentration occurs.

The fact that a lag effect may occur at short eating times does not explain the discrepancy between Overbosch's prediction and our data. The *Tmax* values in Overbosch's work were considerably longer than those obtained in this present study, so the deviation at low *Tmax* values would not have been visible. Closer inspection of Overbosch's original data, however, shows that all the panelists' TI and TR curves were averaged using standard TI evaluation techniques and then analyzed. It is our belief that this process has effectively hidden the pattern we observed. The analogy is that if all the data in Fig. 7 were averaged, the mean value would be very close to 1.0

The developments in analytical technology and methodology are now allowing the study of the basic stimulus-response relationship in substantial detail. The combination of instrumental analysis with sensory analysis is important as are researchers who can assimilate both types of science.

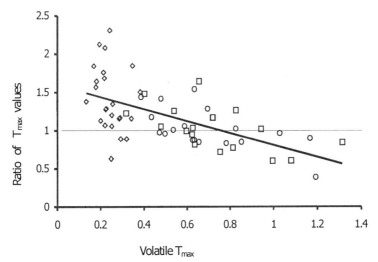

Figure 7. Ratio of sensory and volatile Tmax values plotted against volatile Tmax. If no effect were seen, the ratio should be constant. There is a distinct trend showing deviation from 1.0 at early and at later eating times. The symbols in figure 7 represent three volatiles in a 1 to 8% gelatine-sucrose system.

Interaction between flavor compounds

The fact that flavor compounds interact and change the flavor perceived, is well accepted by flavor researchers although the mechanisms and levels at which it occurs are not fully understood. Physical chemistry tells us that interactions can take place in solution when micelles or other agglomerates form, changing the release characteristics of a particular molecule. The research by Zhao et al (*26*) has demonstrated that a single olfactory receptor has an affinity for a small group of compounds. This has led to the idea that these compounds may compete for the receptor so interaction may also occur at receptor level. Interactions between taste and aroma molecules have also been reported (for overview 27). Ennis (28) has proposed a general receptor model that contains an interaction term and which can be applied to mixtures of compounds. This is an interesting development that would benefit from further experimental testing. Because interaction may affect the interpretation of results, we have generally attempted to use systems where interaction is minimal and we have used single volatiles, which do not interact with the gel matrix used. However, as mentioned in the Introduction, studying just the volatile component of flavor is unlikely to provide all the answers to understanding the complex situation between stimuli and flavor perception. To study some simple interactions, the conductivity and pH in mouth have been monitored using a dental plate (29). A similar system was used by Jack et al (30) to follow conductivity changes in cheese. However, the new system has been tested with a wider range of foods to determine if meaningful results can be obtained, as the release of non-volatiles and their concentration in saliva is bound to be more variable than the concentration of volatiles in a turbulent air flow in-nose.

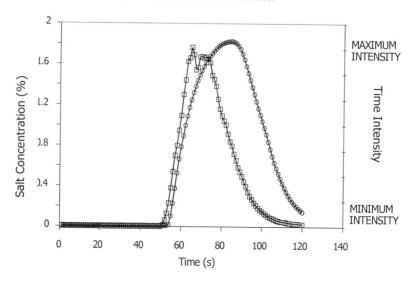

Figure 8. Salt release from mashed potato (square symbols) and the perceived salt intensity (circle symbols) as measured by simultaneous Time Intensity analysis

Fig 8 shows the mean conductivity traces from one person eating three replicate samples of mashed potato along with the simultaneous mean TI trace. The subject was asked to make chewing movements for 50 seconds to establish a baseline, then food was placed in mouth. The conductivity measurements from all three replicate samples showed good reproducibility and were smoothed to remove "noise" from the system. A comparison between the salt Time Release curve and the sensory TI curve (Fig 8) shows some lag between the stimulus and the response. Salt release from low moisture foods like snack foods has also been tested with the conductivity sensor, as has acid release from fruit-flavored gelatin table jellies.

The intention now is to obtain release data for volatiles and non-volatiles along with simultaneous TI for overall flavor and study the relationships for pairs of compound where interactions are thought to occur.

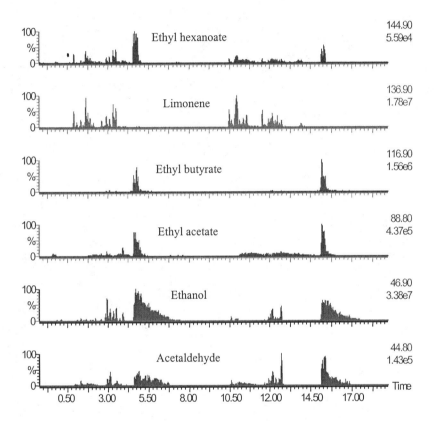

Figure 9. Volatiles released from an orange during preparation and consumption. Values on the right hand side are the m/z values for the ion monitored (top value) and the ion intensity (lower value). The headspace above an intact orange was sampled twice with the first event starting at 0.5 min and the second event at 10.5 min. After sampling the intact fruit, the orange was broken open with the fingers (3.00 and 12 min), and a segment eaten between 5 and 6 min and between 15 and 16 min (volatiles collected from the nose).

Other aspects

The enjoyment of food involves many other factors besides the release of volatiles during the short time of eating. The setting for a meal, the mood of the person eating the meal and the packaging/presentation of the food all contribute. With some foods, however, there are different olfactory signals during the consumption of food and the early signals may serve to build anticipation of the flavor when the food is actually consumed. Coffee is an excellent example where the aroma of brewing coffee has a significant sensory impact, prior to actually drinking the coffee. To gain some idea of the volatile signals, sensed before and during consumption of a food, the API-MS was used to "sniff" an orange during the various stages of preparation and consumption. Fig 9 shows the volatiles in the headspace above an intact, fresh orange, then the change as the orange was peeled, the segments broken open and a segment eaten. It is evident that the volatile profile was different at each stage of preparation and consumption and it raises the question as to whether this type of "aroma experience" is actually important in determining perceived flavor quality. The fact that some coffee producers inject certain volatiles into the headspace of their coffee jars suggests that it does have a significant effect. The question now is whether this approach is useful to enhance the perceived flavor of other food products.

Summary and conclusion

In this paper, some of the reasons for the lack of understanding between stimuli and sensory response have been presented. In our view, it is important to sample volatiles at the appropriate place so that the measured profile will relate better to the sensory response. The significance of the temporal dimension is not yet established and the interaction aspect is at an early stage of investigation. Finally, we should not lose sight of the fact that many factors influence our enjoyment of food and our analyses should not be confined to those that are convenient. To develop this field further needs collaboration between different scientific disciplines.

Acknowledgements

Work on flavor release in the Samworth Flavor Laboratory at the University of Nottingham is funded by the UK Ministry of Agriculture, Fisheries and Food (MAFF), the Biotechnology and Biological Research Council (BBSRC) and a Consortium of industrial companies that include Firmenich (Geneva, Switzerland), Micromass (Altrincham, UK) and Stable Micro Systems (Godalming, UK).

Literature Cited

1. Overbosch, P. *Chem. Senses* **1986**, *11*, 315-329.
2. Stewart, G.F. *Food Technol.* **1963** *17*, 5 only.
3. Williams, A. *Food Qual. Pref.* **1994** *5*, 3-16.

4. Hoppe, K. Psychophysical relationships: Steven's and/or Fechners Law? In *Flavor Perception* Kruse, H-P.; Rothe, M, Eds.; Eigenverlag Universitat Potsdam **1997** pp27-34.

5. Lawless, H.T. Olfactory Psychophysics In *Tasting and Smelling* Beauchamp, G.K.; Bartoshuk, L. Eds.; Academic Press. Second edition **1997** pp.125-174.

6. Halpern,B.P. Psychophysics of Taste In *Tasting and Smelling* Beauchamp, G.K.; Bartoshuk, L. Eds.; Academic Press. Second edition **1997** pp.77-123.

7. Taylor, A.J. *Crit. Rev. Food Sci. Nutr.* **1996** *36*, 765-784.

8. Taylor, A.J., Linforth, R.S.T., Flavor release from foods: Recent developments In *Flavor Perception* Kruse, H-P.; Rothe, M. Eds.; Eigenverlag Universitat Potsdam **1997** pp131-142

9. Laing, D.G. *Food Qual Pref,* **1994** *5*, 75-80.

10. Laing, D.G.; Jinks A., *Trends Food Sci. Technol.* **1996** *7*, 387-389.

11. Overbosch, P.; Afterof, W.G.M.; Haring P.G.M. *Food Rev. Int.* **1991** *7*, 137-184.

12. Taylor, A.J.; Linforth, R.S.T., *Trends Food Sci. Technol.* **1996** *7*, 444-448.

13. Linforth, R.S.T.; Taylor, A.J. *Food Chem.* **1993** *48,* 121-126.

14. Ingham, K.E.; Linforth, R.S.T.; Taylor, A.J. *Lebensmitt. Wiss. Technol.* **1995** *28*, 105-110.

15. Roozen, J.P.; Legger-Huysman, A. Sensory analysis and oral vapour gas chromatography of chocolate flakes, In *Aroma: perception, formation and evaluation*, Rothe, M.; Kruse, H-P. Eds.;, Eigenverlag Deutsches Institut fur Ernahrungsforschung, Berlin **1995** pp. 627-632.

16. Delahunty, C.M.; Piggott, J.R.; Conner, J.M.; Paterson, A. *J. Sci. Food Agric.* **1996** *71*, 273-281.

17. Soeting, W.J.; Heidema, J. *Chem. Senses* **1988** *13*, 607-617.

18. Linforth, R.S.T.; Taylor, A.J. "Apparatus and methods for the analysis of trace constituents in gases", *European Patent Application* EP 0819 937 A2 1998

19. Benoit, F.M.; Davidson, W.R.; Lovett, A.M.; Nacson, S.; Ngo, A. *Anal. Chem.* **1983** *55*, 805-807.

20. Devos, M.; Patte, F.; Rouault, J.; Laffort, P.; Van Gemert, L.J. *Standardized human olfactory thresholds* IRL Press, Oxford, 1990.

21. Brauss, MS.; Linforth, R.S.T.; Cayeux, I.; Harvey, B.; Taylor, A.J. *J. Agric. Food Chem.* In press.

22. Overbosch, P.; Van den Enden, J.C.; Keur, B.M. *Chem. Senses* **1986** *11*, 331-338.

23. Overbosch, P.; De Jong, S. *Physiol. Behavior* **1989** *45*, 607-613.

24. Overbosch, P.; De Wijk, R.; De Jonge, Th.J.R.; Koster, E.P. *Physiol. Behavior* **1989** *45*, 615-626.

25. Baek, I.N.; Linforth, R.S.T.; Blake, A.; Taylor, A.J. *Chem. Senses* **1998,** in press.

26. Zhao, H.Q.; Ivic, L.; Otaki, J.M.; Hashimoto, M.; Mikoshiba, K.; Firestein, S. *Science*, **1998** *279*, 237-242.

27. Noble, A.C., "Taste-Aroma interactions", *Trends Food. Sci. Technol.* **1996** *7*, 439-443.

28. Ennis, D.M. *Food Chem* **1996** *56*, 329-335.

29. Davidson, J.M.; Linforth, R.S.T.; Taylor, A.J. *J. Agric. Food Chem.* **1998** *46*, 5210-5214.

30. Jack, F.R.; Piggott, J.R.; Paterson A. *J. Food Sci.* **1995** *60*, 213-217.

INDEXES

Author Index

Subject Index

172

Highlights from ACS Books

Desk Reference of Functional Polymers: Syntheses and Applications
Reza Arshady, Editor
832 pages, clothbound, ISBN 0–8412–3469–8

Chemical Engineering for Chemists
Richard G. Griskey
352 pages, clothbound, ISBN 0–8412–2215–0

Controlled Drug Delivery: Challenges and Strategies
Kinam Park, Editor
720 pages, clothbound, ISBN 0–8412–3470–1

Chemistry Today and Tomorrow: The Central, Useful, and Creative Science
Ronald Breslow
144 pages, paperbound, ISBN 0–8412–3460–4·

A Practical Guide to Combinatorial Chemistry
Anthony W. Czarnik and Sheila H. DeWitt
462 pages, clothbound, ISBN 0–8412–3485–X

Chiral Separations: Applications and Technology
Satinder Ahuja, Editor
368 pages, clothbound, ISBN 0–8412–3407–8

Molecular Diversity and Combinatorial Chemistry: Libraries and Drug Discovery
Irwin M. Chaiken and Kim D. Janda, Editors
336 pages, clothbound, ISBN 0–8412–3450–7

A Lifetime of Synergy with Theory and Experiment
Andrew Streitwieser, Jr.
320 pages, clothbound, ISBN 0–8412–1836–6

For further information contact:
Order Department
Oxford University Press
2001 Evans Road
Cary, NC 27513
Phone: 1-800-445-9714 or 919-677-0977
Fax: 919-677-1303